# Archaeology and Coastal Change

Edited by F. H. Thompson

Being the papers presented at meetings in London and Manchester on 27th October and 5th November, 1977

Occasional Paper (New Series) I

THE SOCIETY OF ANTIQUARIES OF LONDON

Burlington House, Piccadilly, London W1V 0HS
1980

ISBN 0 500 99031 X



TYPESET BY BISHOPSGATE PRESS
PRINTED BY WHITSTABLE LITHO

# *Contents*


*Editorial Note* iv

*Chairman's Introduction* v

On Sea-Level Changes by *C. E. Everard* 1

Caerleon and the Gwent Levels in Early Historic Times by *G. C. Boon* 24

The Evolution of Romney Marsh: a Preliminary Statement by *B. W. Cunliffe* 37

Iron Age and Roman Coasts around the Wash by *B. B. Simmons* 56

Theories of Coastal Change in North-West England by *M. J. Tooley* 74

Archaeology and Coastal Change in the North-West by *G. D. B. Jones* 87

Possible Evidence for Sea-Level Change in the Somerset Levels by *F. A. Hibbert* 103

Archaeology and Coastal Change in the Netherlands by *L. P. Louwe Kooijmans* 106

Post-glacial Environmental Change and Man in the Thames Estuary: a Synopsis by *R. J. Devoy* 134

*Index* 149

# *Editorial Note*

By one of those coincidences in the world of learning, two one-day seminars on the theme of archaeology and coastal change took place independently and almost simultaneously in the autumn of 1977. The first was organized by the Society of Antiquaries as a one-day symposium, *Archaeology and Coastal Change in Southern Britain*, under the chairmanship of Professor G. W. Dimbleby, F.S.A., and was held in the meeting room of the Geological Society on 27th October. The second was organized by the Extra Mural Department of the University of Manchester as a one-day school, *Archaeology and Coastal Change,* under the direction of Professor G. D. B. Jones, F.S.A. and was held in the University Roscoe Building on 5th November.

Subsequently, when the possibility of publication was considered, there was ready agreement by both sides that they should appear together in one volume, and we are grateful to Professor Jones and his colleagues for their whole-hearted collaboration. We have not attempted complete integration of the two sets of papers, partly to retain the identity of the two meetings: the London contributions by Boon, Cunliffe, Everard and Simmons, occupy pp. 1–73, and the Manchester ones, by Tooley, Jones, Hibbert, Kooijmans and Devoy, pp. 74–148. We also felt that publication should be achieved fairly rapidly and economically, which precluded the addition of this publication to the Research Report series. Instead, we have reverted to the short-lived series of *Occasional Papers*, of which the Society published two in 1943 and 1945, and we are distinguishing this and its followers (of which we hope there will be many) by the addition of the suffix '(New Series)'.

*Society of Antiquaries*
*August 1979*

F. H. THOMPSON
*General Secretary and Editor*

R. R. COLLINS
*Assistant Editor*

# *Introduction*

*by*

*Professor G. W. Dimbleby, F.S.A.*

The research importance of the theme of 'Archaeology and Coastal Changc' is shown by the fact that two conferences should have been independently organised on the theme, and that each was able to call on a quite different team of very competent speakers. As a subject for research, coastal change offers one of the best examples of interdisciplinary integration between archaeology and the physical and biological sciences. It is encouraging to see the environmental scientists using archaeological evidence (e.g. the papers by Tooley and Devoy) and the archaeologists integrating the scientific evidence with their own observations (Boon, Cunliffe, Simmons, Jones and Louwe Kooijmans). The only paper that does not incorporate archaeology with the environmental evidence is that by Everard. This was designed to present the state of our knowledge of sea-level changes, which Everard does most comprehensively, setting the British scene in the context of global changes. This paper therefore provides the fundamental background to the other papers presented at the two conferences.

Though the two conferences took place at widely different venues, neither had a strong local motif, so that, taken together, they cover the west, south and east coasts of England and Wales, with an important contribution on Holland. It would be valuable to have similar research presented from those parts of the British Isles more remote from the continental mainland, particularly Scotland and Ireland. Relevant research is being carried out in these areas and this should complement the evidence presented in this volume.

Apart from the sea-level data presented by Everard, Tooley and Devoy, and to a lesser extent by some of the other authors, there is a major element from the biological fields in these papers, and this makes it possible to see coastal change not just in terms of sea-level but also in terms of the resultant landscape, a distinction of great importance to archaeology. Organic layers such as peats not only tell their own story by their very presence but they are a source of organic remains like pollen from which the local vegetation can be inferred. They also provide ideal material for radiocarbon dating, on which several of the papers in this collection lean heavily. Other plant remains occur in other contexts; diatoms in particular have proved a valuable source of information about salinity (Devoy, Louwe Kooijmans). Animal remains are less widely occurring but on occasion foraminifera and molluscs can provide direct environmental evidence (e.g. Tooley, p. 78; Boon, p. 30), whilst peat beds may yield vertebrate bones as well as plant remains (e.g. Devoy, p. 144). However, archaeological sites are the richest source of animal remains, especially of fish and molluscs, which may indicate the nature of the coasts and estuaries being exploited by man. The sites quoted by Louwe Kooijmans have yielded some significant occurrences such as the abundant remains of sturgeon at Hazendonk (p. 119), and these and several other sites mentioned among these papers have also produced botanical material indicative of both the ecological and cultural environs of such coastal sites.

In several of the papers (Simmons, Cunliffe, Boon, Jones and Louwe Kooijmans) the positions of the archaeological sites are related to the alignment of the contemporary coastline, and indeed the position of the coastline can sometimes be inferred, in the absence of other evidence, from the location of the sites. Jones, Simmons, Boon and Cunliffe all discuss the deposition of alluvium since Roman times, which has isolated Roman coastal sites from the present coastline. It was outside the scope of these conferences, but nevertheless would be of both archaeological and ecological interest, to consider the possible relationship between the intensity of land use inland and the concomitant accumulation of silt and alluvium in estuaries and along the coast. There are several instances quoted in these papers of the effects of human activity in medieval and later times through the construction of dykes, levees and drainage works (Boon, Cunliffe, Devoy), but I suspect that long before this agriculture was adding considerably to our coastal and off-shore deposits. It would be a logical extension of the theme of these conferences to study a whole river catchment and attempt to relate the land use in it to the coastal conditions adjacent to its mouth. Among these papers good candidates are to be found from North-West England, the Wash, and the South Coast.

# On Sea-Level Changes

C. E. Everard

Attempts to measure and explain changes in the height of the sea in both time and space lie largely in the province of the geoscientist. The archaeologist has a direct interest in the magnitude and dating of late Pleistocene and Holocene[1] sea-level changes in southern Britain in so far as they expanded or shrank the available living space of the coastal and estuarine populations.

This paper attempts no more than a brief review of some of the problems underlying the measurement of contemporary sea level and its fluctuations and their implications for the interpretation of past sea levels. Akeroyd (1972) has warned against ' . . . a too ready and uncritical acceptance of the evidence for apparent changes in relative land and sea level . . . ' and her paper and thesis (Akeroyd 1966) must be regarded as basic reading for anyone embarking on a study of archaeology and sea level changes. The literature on sea levels is enormous in quantity and grows exponentially: it is scattered among the works of astronomers, geophysicists, geologists, geomorphologists, hydrographers, oceanographers, climatologists, biologists, archaeologists, historians, land surveyors, civil engineers etc. The list amply emphasises the very complex interdependence of the dynamics of the solid earth, the world ocean and the atmosphere. Tidal oscillations are superimposed upon more persistent changes in sea level and it is 'on the coast that these interrelationships are most forcibly and dramatically evident' (Walcott 1975).

Figure 1 indicates in a general way the amplitude spectrum of periodic water level variations for the oceans, ranging from waves to geological inundations of continental proportions. The 5 m span has been added, which is claimed by some to be the range of mean sea level fluctuations over the past 6000 years in southern Britain. Although well clear of the amplitude of the main climatically induced changes (i.e. glaciations) it is uncomfortably close to the larger tidal amplitudes and it is, of course, in tidal coastal locations that many relevant archaeological sites lie and in which many contemporary sea level movements are recorded.

According to Geyhl and Streif (1970) the hypothesis that different sea level heights in coastal regions are represented as continental and marine phases is about 40 years old, the mineral, clastic sediments being interpreted as marine and the interbedded peat as continental. The regional significance of such vertical

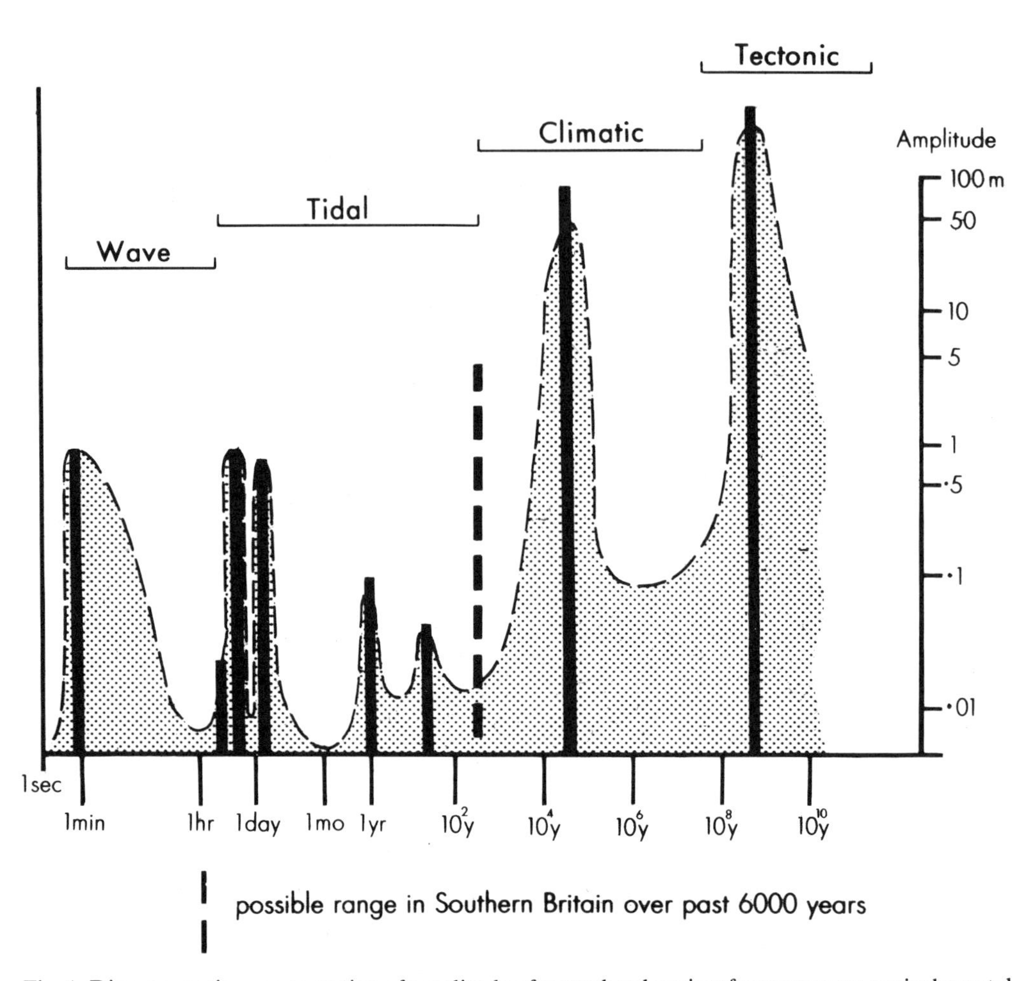


Fig. 1. Diagrammatic representation of amplitude of water level against frequency at a typical coastal location (Walcott 1975).

sequences, as opposed to purely local movements, remains uncertain. The presence or survival of peat may be affected by tectonic movements, storm erosion, halokinetic effects etc. Precise dating may allow stratigraphic correlation and from this the horizontal displacement of shorelines may be assessed, but differential compaction of sediments makes the quantitative correlative of the level of the sea difficult.

What is recorded at any one site by instrumental, stratigraphic, or other means is, of course, a *relative* sea-level change compounded of two movements, one by the sea, the other by the land (either may of course be zero), as follows:

(i) an absolute change in ocean level, a eustatic change which, since all oceans and their marginal seas are inter-connected, will be world wide (but see p. 8);

(ii) an absolute change in the height of the land, i.e. a tectonic movement, on a local, regional or even continental scale, relative to the centre of the earth.

It is with the greatest difficulty that these two components can be separated and quantified.

Much intensive research has, since the 1960s, been directed to eliminating tectonic influences (ii, above) so that a world-wide eustatic curve of sea level change in the late Pleistocene and Holocene could be constructed. Figure 2 presents some of the diverse results.[2] More recently this whole concept has been challenged and it may be that the eustatic hypothesis is applicable only at the regional level.

A very simplistic explanation of relative mean sea level movements is given in fig. 3. Each such movement measured over a period of, say, decades or centuries, is the algebraic resultant of tectonic movement and eustatic change:

$$\text{MSL} = \text{TM} + \text{EC}$$

TM and EC can move in either the same or opposite senses and at different rates, so that the 10 combinations shown in fig. 3a are possible, but to an observer on the coast only three trends can be recognised: a sea level that is stable, rising (positive change) or falling (negative change) (fig. 3b). Quantitatively the rates of change may vary over quite short distances (p. 21).

It follows from the above that three steps are necessary to determine contemporary sea-level movements and to assess evidence of past movements:

(i) the definition of average or mean sea level
(ii) the measurement of mean sea level
(iii) the measurement of crustal movements.

Gordon and Suthons (1963) noted that there was a tendency until recent years for land surveyors to assume a stable sea level against which they could measure crustal change and for hydrographers to assume a stable land against which they could compare sea level movements. Stability of neither can be assumed.

## *Mean sea level*

Johnson (1929) stated that tidal studies have twin objectives: (i) to determine the *form* of the mean sea level surface, (ii) to make comparative studies of sea level changes. Most contemporary sea level changes (i.e. within the past century) lie within or close to the tidal range and can only be studied by examining the average or mean level of the sea. It is possible that mean high tide, or mean high spring tide, would be a better datum for archaeological purposes but mean sea level itself is fraught with enough problems and many of the following comments would apply to high tide means as well. Jardine (1976) has argued for the use of mean tide level rather than mean sea level, as the former is related to the tidal range, which in turn is modified by coastal configuration.

There now exists the International Geological Correlation Programme (I.G.C.P.) designed to run from 1974 to 1982 by which it is hoped to obtain, by processing existing data and collecting new heights and dates, the trend of mean

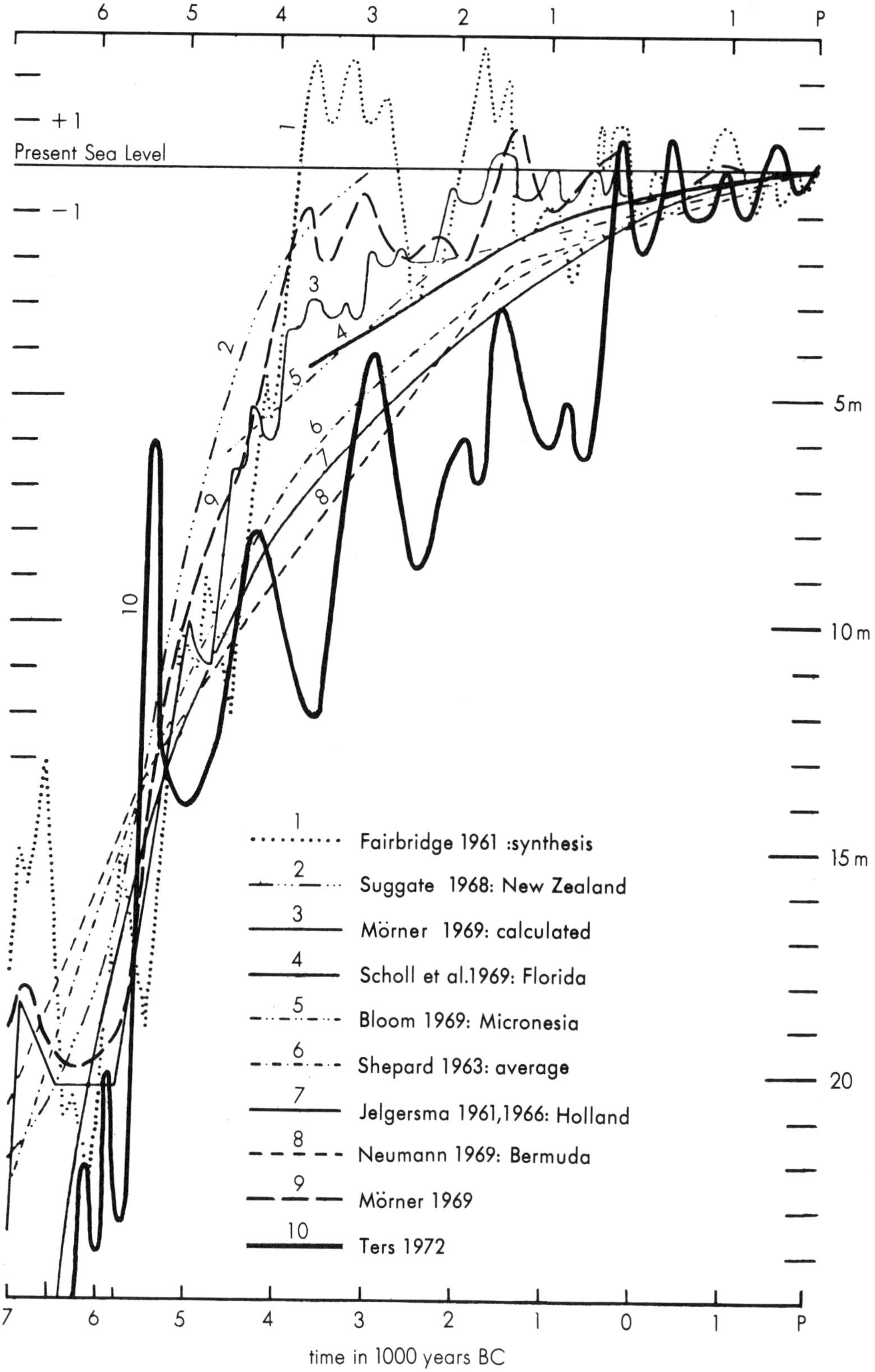
6
5
4
3
2
1
1
P
+1
Present Sea Level
−1
5m
10m
15m
20
1
2
3
4
5
6
7
8
9
10
1 Fairbridge 1961 :synthesis
2 Suggate 1968: New Zealand
3 Mörner 1969: calculated
4 Scholl et al.1969: Florida
5 Bloom 1969: Micronesia
6 Shepard 1963: average
7 Jelgersma 1961,1966: Holland
8 Neumann 1969: Bermuda
9 Mörner 1969
10 Ters 1972
7
6
5
4
3
2
1
0
1
P
time in 1000 years BC

Fig. 2. Reconstructions of sea level changes during the past 8000 years (Ters 1973; Mörner 1976).

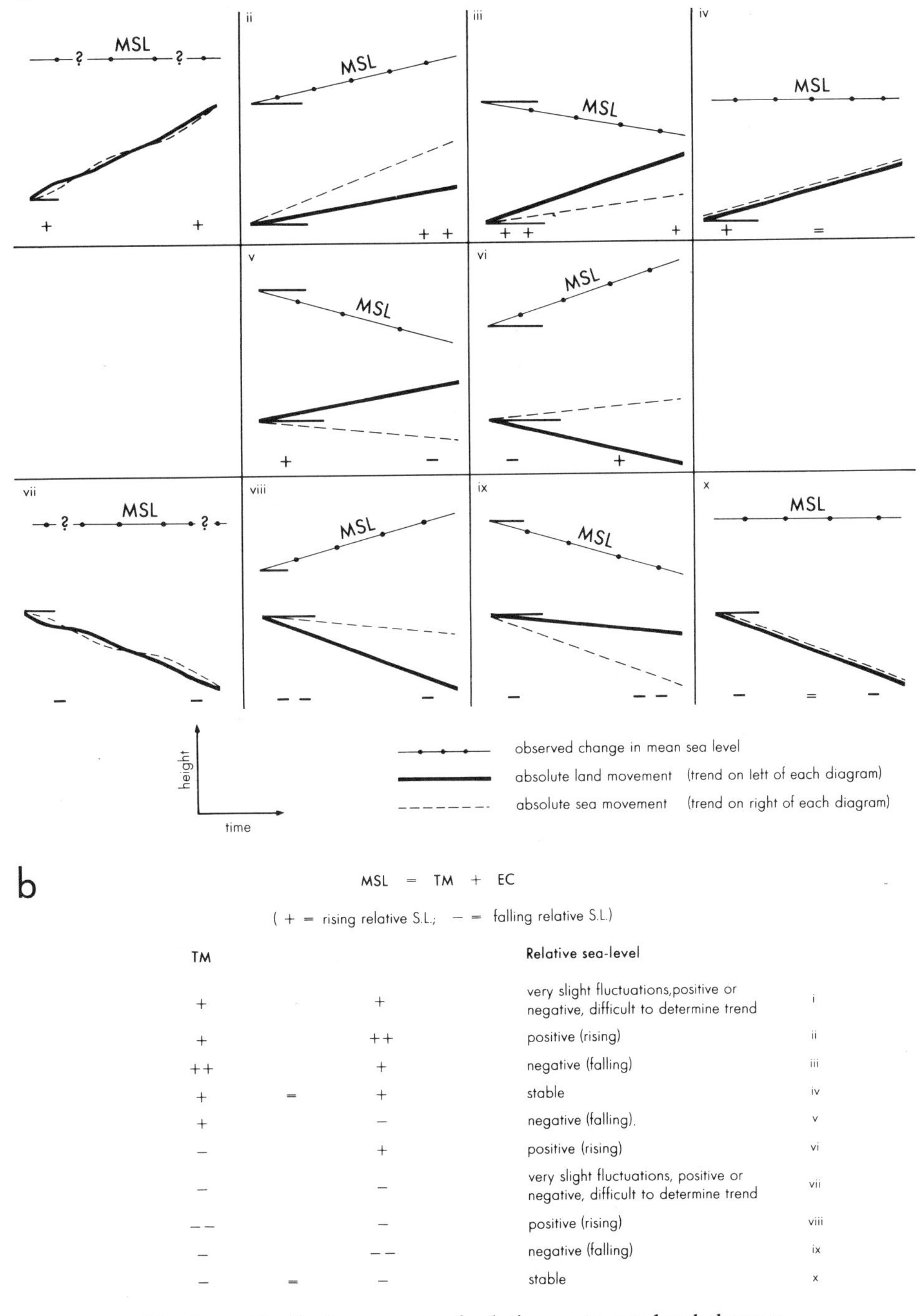


| TM | | | Relative sea-level | |
|---|---|---|---|---|
| + | | + | very slight fluctuations, positive or negative, difficult to determine trend | i |
| + | | ++ | positive (rising) | ii |
| ++ | | + | negative (falling) | iii |
| + | = | + | stable | iv |
| + | | − | negative (falling). | v |
| − | | + | positive (rising) | vi |
| − | | − | very slight fluctuations, positive or negative, difficult to determine trend | vii |
| −− | | − | positive (rising) | viii |
| − | | −− | negative (falling) | ix |
| − | = | − | stable | x |

Fig. 3a and b. Basic concepts of relative mean sea level changes.

sea level 'during the last de-glacial hemicycle and continuing to the present . . . as a basis for mathematical modelling and the development of predictive modes' (Tooley 1976).

Wemelsfelder (1971) identifies two quite different interpretations of mean sea level: (i) a period (e.g. monthly) mean at a place, which is fairly easily measured but not necessarily constant. Monthly averages in the North Atlantic range from +10 cm (3.93 in.) in October to −10 cm (−3.93 in.) in February against a nominal zero and long term cyclic trends are also present. Such means are used for local chart datums, island surveys, etc.; (ii) physical mean sea level, the presumed world-wide level when all the disturbing influences have been filtered out. This is the generally accepted concept of mean sea level but it is not an equipotential surface and was on theoretical grounds believed to have inherent irregularities. In recent years this real ocean surface, called the geoid, has been shown by satellite geodesy to have humps and depressions of several tens of meters (the maximum 'relief' reaches 180 m (590.6 ft.) in the S.W. Pacific (Mörner 1976)). More recent work on the mean sea level concept is reviewed by Everard (in press).

An additional point, based initially on surveys in the New York harbour area (Johnson 1929), is that ' . . . along an irregular coast the mean sea level surface is an irregularly warped plane which is extremely sensitive to alteration in the form of the shore, the depths of the tidal channels and other shore features . . . changes in these features . . . cause local fluctuations in measured mean sea level ranging from fractions of an inch to . . . one or two feet.'

It has been suggested that mean sea level is an illusory concept, but most research at present seeks to identify and even to quantify the factors which influence contemporary sea levels, that we may the better interpret evidence of earlier ones. A thematic approach (cf. Lisitzin 1974) might examine the following interacting factors:

| | |
|---|---|
| Astronomical: | e.g. nodal tides (18.6 year period) etc. |
| Oceanographic: | sea water temperature, salinity etc. |
| Climatological: | proportion of water in solid, liquid and gaseous states |
| Meteorological: | pressure, winds etc. |
| Hydrological: | fresh water discharge |
| Technical: | tide gauge design, observational practice etc. |

Alternatively the factors may be reviewed on a spatial basis (cf. Wemelsfelder 1971) adopting an arbitary division into global, regional and local, together with instrumental factors, the pattern followed in this paper.

## *Global Influences*

1. Ellipticity. The shape of the geoid (sea level surface) is affected by the earth's rotation. The latter is related in part to the coupling of the earth's core/mantle interface and its changes must alter the configuration of the geoid, both vertically and horizontally. For these Mörner (*op. cit.*) has introduced the term geoidal eustasy, which can cause marine transgressions and regressions that have no

climatic or tectonic involvement. He argues that 'by geoidal eustasy the ocean level may change quite differently over the globe and even change sign'.

The rate of rotation, and hence geoid shape, may also be affected by the addition or removal or water from the oceans (Munk and MacDonald 1960). It has been calculated that a +10 cm (3.93 in.) sea level change in 50 years would decrease rotation by one part in $10^8$ (Fairbridge and Krebs 1962), which would presumably decrease the tendency for the polar flattening and equatorial bulging of the geoid.

2. Astronomical influence. Currie (1976) has shown, for example, that certain tide gauge records of the late nineteenth and twentieth centuries show 18.6 year (lunar), 10.9 year (solar) and 6.3 year (unknown origin) cycles, with amplitudes of 9 mm (.25 in.), 9 m (10 yd.) and 14 mm (.5 in.) respectively.

3. Quantity of water.

(a) It has been argued on the one hand that the total amount of water in the earth-atmosphere system has remained constant since the end of the Pre-Cambrian (*c.* 600 m.y. ago) and on the other that it has increased during this time by the addition of juvenile water of volcanic origin (Turekian 1976). As the past 6000 years have not been exceptionally volcanic this factor will not be pursued in the present context.

(b) The distribution of water within the various parts of the hydrological cycle will vary *inter alia* with the average temperature of the atmosphere: a warm atmosphere will hold more water vapour than a cold one (global mean temperature fell in the period 1950–73), but as the present atmosphere contains only 0.035% of the non-oceanic water in the cycle and as the latter is under 3% of the total, temperature fluctuations of the atmosphere itself are not significant in the present context.

(c) Of more significance is the amount of water in the solid state. The *c.* 100 m sea level rise of the Flandrian transgression, related to the melting of glacier ice, is an example of glacio-eustatism. In the past 6000 years the greater part of the world's ice has been in the Antarctic and Greenland ice sheets, the rest in small ice caps and mountain glaciers. If the present mountain glaciers of Europe, Asia and North American were to melt, sea level would rise only by 60 cm (23.6 in.) and even at their Pleistocene maxima the rise would have been only 3 m (9.84 ft.) (Flint 1971).

Denton and Karlen (1973) claim that there is no systematic relationship between the known fluctuations of non-Polar ice and Holocene sea-levels and volume changes in the enormous Antarctic and Greenland ice caps clearly have a much greater influence on Holocene sea levels. A one-metre sea level change is equal to only 23 m (75.46 ft.) of ice spread uniformly over Antarctica (Walcott 1975) but it seems likely that Holocene sea levels have been much more influenced by growth or loss of ice at the margins of the ice sheets. Paterson (1972) has calculated that a one-metre change in ocean level would result from an average horizontal shift of 10.6 km (6.5 miles) in the edge of the Antarctic ice sheet: for the smaller Greenland sheet it would be 100 km (62 miles). Radiocarbon dates of 6000 bp are known only 20 km (13 miles) from the edge of the Greenland ice, which suggests little change in position that could influence the sea level and there is no conclusive evidence as yet of the trend of the two ice cap margins over the past few thousand years.

(d) A rise of 1°F in a water column 183 m (600 ft.) deep will expand it by 2.54 cm (1 in.). A salinity decrease from 35 to 34.9‰ raises local sea level by 1.9 cm (.75 in.). Hoinkes (1960) claims that an increase of average sea water of only 0.007°C per year would offset a eustatic fall of 2.9 mm (.0125 in.) per year caused by the (presumed) growth of the Antarctic ice cap and would, in fact, produce a net overall rise in sea level of 1.1 mm (.0375 in.) per year (which is a figure accepted by many workers as the current eustatic rise).

4. World ocean currents influenced by wind drift, density contrasts and the Coriolis effect modify the sea level but are not relevant in the present context (see Wemelsfelder 1971, p. 121).

5. The ocean basins undergo continuous deformation by a variety of forces. Sea level is therefore altered and for this Mörner has used the term tectonic-eustasy.

(a) Tectonic-plate movements. The Atlantic basin, for example, is widening at rates of the order of 2–4 cm/year (.75–1.5 in.) at present. In addition, the ocean floor is rising at constructive plate margins (the so-called mid-ocean ridges), progressively subsiding as it is moved away from them and deeply subsiding at the subduction zone oceanic trenches, particularly those fringing the Pacific, resulting in a continual alteration in ocean basin capacity.

(b) Volcanic activity away from plate margins, e.g. the Darwin Rise (central Pacific Ocean) leads to crustal loading and isostatic depression to the extent that many major oceanic volcanoes are surrounded by depressed 'moats'.

(c) A eustatic sea level change, however caused, is a re-distribution of load over the face of the globe, a load which in turn deforms the global surface (elastic deformation), i.e. eustatic and tectonic movements become interdependent. Walcott (1975) and Farrell and Clark (1976) have shown conclusively that we cannot partition crustal and sea level movements because isostatic re-adjustments affect the whole earth (see Pirazzoli 1976). Walcott (1975) calculated that to ignore the re-distribution of water from melting ice sheets on crustal deformation may lead to a 30% error in assessing the final 'eustatic' change. Farrell and Clark (1976) dispose of the usual assumption that sea level changes are uniform over the ocean basin when ice and water masses are re-arranged. As an example they show that the melting of ice in the Laurentide and Fennoscandian sheets, equivalent to a uniform 100 m (328 ft.) sea level rise, would lead to an instantaneous 120 m (394 ft.) rise in the Pacific, but one of under 100 m (328 ft.) in the North Atlantic. One thousand years after the melting further crustal adjustment ('relaxation time') causes more sea level changes (fig. 4a and b). They make a further point, often overlooked: 'the gravitational attraction of an ice mass upon a nearby ocean tends to hold sea level high in the vicinity of the ice. This extra load near the ice may have a significant influence on postglacial isostatic adjustment.' The authors claim that there is no need to define a universal eustatic sea level curve and argue that field observations should not be 'corrected' for eustatic changes. Their paper does suggest however, that following the melting of the Pleistocene ice sheets in the Northern Hemi-

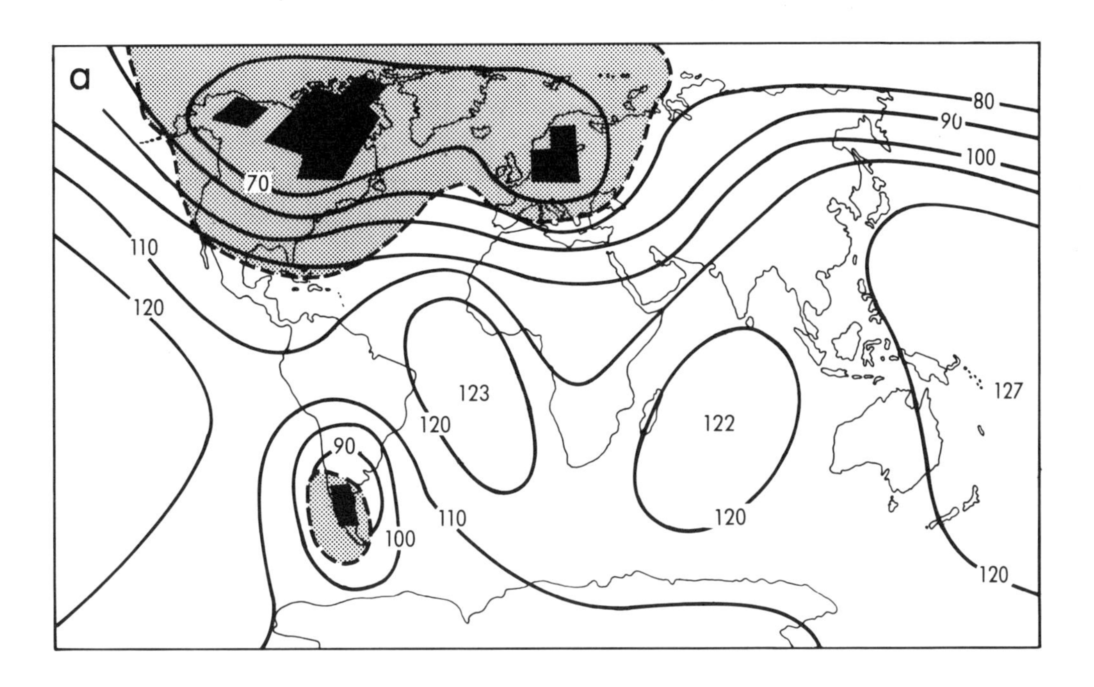


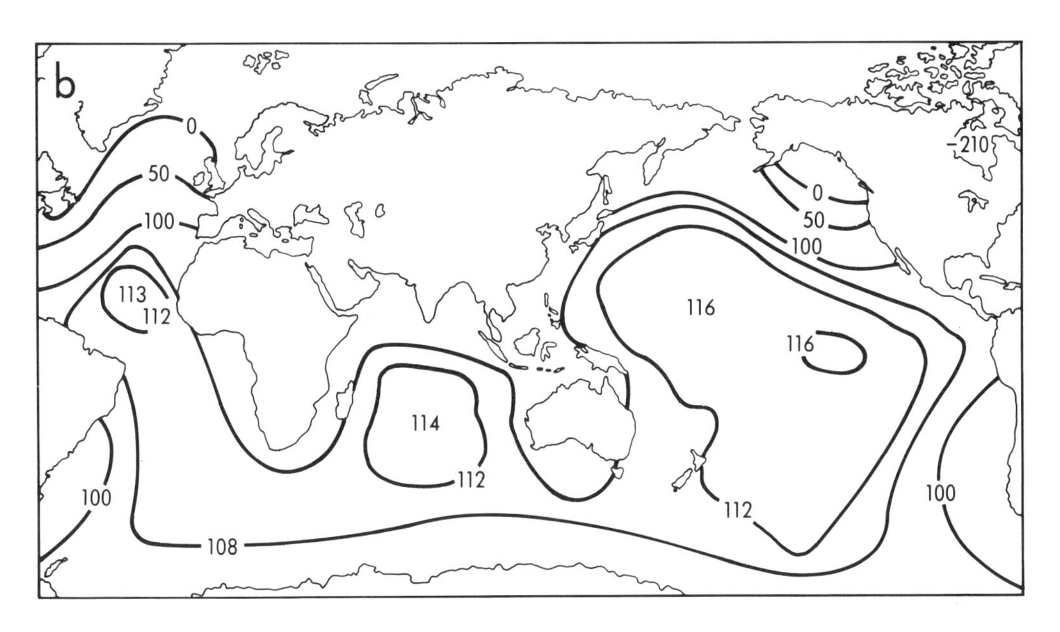


Fig. 4a. Vertical displacement caused by elastic deformation of the Gutenburg earth by a rise in sea level derived from melted ice in the black areas. Contours are sea level change expressed as a percentage of the eustatic (oceanwide average) change in sea level (from Walcott 1972). b. Total percentage change in sea level after melting 1 m from all [Pleistocene] ice masses, excluding Antarctica, and allowing the earth to relax for 1000 years (Farrell and Clark 1976).

sphere a relative north to south sea level rise might be expected across Britain.

6. Further global load redistribution is related to continental erosion and continental terrace/ocean floor sedimentation. Continuous sedimentation has affected the Atlantic and Indian Oceans, and to a less extent the Pacific, during the Quaternary. It has been estimated (Turekian 1976) that the average rate of lowering of the continents is *c.* 6 cm/1000 years (2.25 in.), the load being transferred to the oceans but, of course, compensating isostatic movements develop. Umbgrove (1947) postulated the 'continental flexure', by which the coastal plains rose as the continental terrace sank, while King (1950) envisages a pulsating movement, reflected in widespread continental planation. Whether continuous or spasmodic, the continent/ocean adjustment is ongoing.

### *Regional Influences*

Global factors tend ultimately to have expression as eustatic sea level changes. The most important regional factors are probably crustal disturbances.

7. An increase of 1 mb in atmospheric pressure depresses sea level by 0.99 cm (.375 in.) and vice versa. A shift of average pressure over time, linked with (8) below might appear as a sea level change.

8. Prevailing winds, in terms of monthly resultants, may reflect a long term climatic change, such as the decreasing prevalence of westerlies over Britain in recent decades (Lamb 1972) and induce a sea level change related to (7) or (10).

9. Seasonal changes in temperature and salinity cause the North Atlantic to be *c.* 25 cm (9.75 in.) higher in summer and autumn than in spring (2d, above) but long trend influences on mean sea level are complex. Since 1876 the average North Atlantic temperatures decreased in the first few years, rose, then fell again to a minimum *c.* 1920, rising again to a maximum in the 1930s (or 1950s in eastern parts), since when it has been falling (Perry and Walker 1977).

10. Any body of water moving across the earth's surface is deflected by the Coriolis effect (to the right of motion in the northern hemisphere) and also by centrifugal forces. A 4–8 cm (1.5–3 in.) north to south rise across the English Channel has been related to the former (Cartwright and Crease 1963)[3] and in 1970 d'Hieres and Le Provost used a rotating model of the English Channel to simulate the Coriolis effect on its currents. They found that mean sea level varied with the strength of the tidal currents, especially near straits and islands where there were curving, accelerating currents. Their detailed prediction for the Isle of Wight area (fig. 5) suggests that important local mean sea level variations may exist over very short distances in areas of strong tidal currents. It is obvious that minor changes in shoreline configuration caused by changes in 'geoid' sea level could have major consequences on the contours derived by d'Hieres and Le Provost.

11. Although not an 'active' crustal area on a world scale, Britain is by no means tectonically stable.

(a) It can be shown (King 1954, Brown 1960) that the foundation of southern Britain is a series of raised or depressed blocks of Palaeozoic and earlier

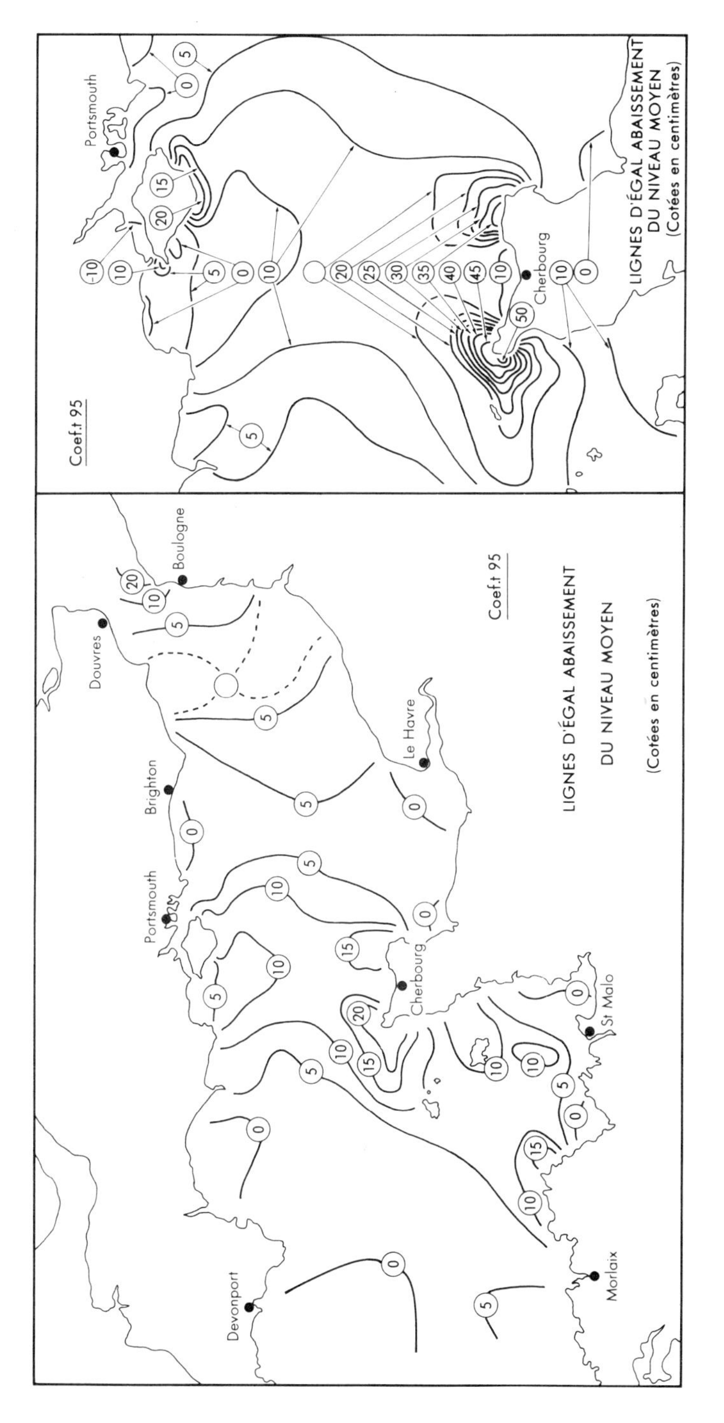


Fig. 5. Predicted variations in mean sea level in the English Channel caused by the Coriolis effect (d'Hieres and Le Provost 1970).

rocks (fig. 6) that originated during earth movements in Permo-Triassic times. The concept is readily extended to the rest of Britain and adjacent Europe. Differential movement between the blocks and differential compaction of their covering sediments must be allowed for in any mean sea level studies. For example, upper Pliocene beds lying at 180 m (591 ft.) O.D. on the North Downs have probable equivalents at 0 m O.D. in East Anglia, 35 m (115 ft.) O.D. in west Cornwall and below sea level in the western English Channel (cf. West 1972).

(b) Recent geological studies (Delany 1971) of the western approaches to the English Channel indicate 6–8 marine transgressions in the Mesozoic–Cainozoic, demonstrating an unstable continental margin to the west of Britain, which must contribute to west–east warping in southern Britain (cf. Churchill 1965).

(c) Within even such a small area as the West Country arguments have been presented for isostatic buoyancy in Devon and Cornwall (Bott and Scott 1964) and depression in Lundy (Brooks 1973).

(d) In the present context the sedimentary load of the Rhine delta presumably contributes to the apparent relative downward trend of the crust in the southern North Sea. It is unfortunate that the world's longest continuous record of mean sea level change should be that of the Amsterdam tide gauge (records have been kept since 1682). Amsterdam is the zero for the European levelling network (p. 17).

(e) Celsius, in 1743, was one of the first to record the rise of Fennoscandia following the melting of the Pleistocene ice sheets, a movement today known as glacio-isostasy. The rates of uplift are not uniform but decrease outwards from a maximum near the head of the Gulf of Bothnia and various authors have suggested that beyond a 'hinge line' there is a complementary subsiding peripheral area, which would include southern Britain, possibly sinking at $^1/_{10}$ the maximum uplift rate (Walcott 1975). This is the concept of 'isostatic rebound', the peripheral zone rising during a glacial because of outward displacement of subcrustal matter by the weight of an ice sheet and then sinking when the subcrustal flow reverses to compensate for the isostatic rise of the now deglaciated central area. In 1972 Walcott suggested that, as the majority of observations on which the Holocene eustatic sea level changes are based lie in these areas of sinking, the rising trend deduced from them was misleading, and he proceeded to make a case for a world wide downward eustatic change during this period. To any isostatic rebound movements in southern Britain must, of course, be added the crustal complexities mentioned under (11a) above.

12. The last regional factor we shall discuss is that of storm surges. These have only recently received detailed study and so the historical extension of their identification and importance is very limited.

The 1953 North Sea storm surge, in which tides reached *c.* 2.4 m (7.9 ft.) above prediction, is well documented and one relevant result was the establishment of a British tide gauge system (fig. 7).

Records of east coasts floods go back to at least 1099 and the surge maxima have increased in height through time, but not until 1929 was the wave-like progression

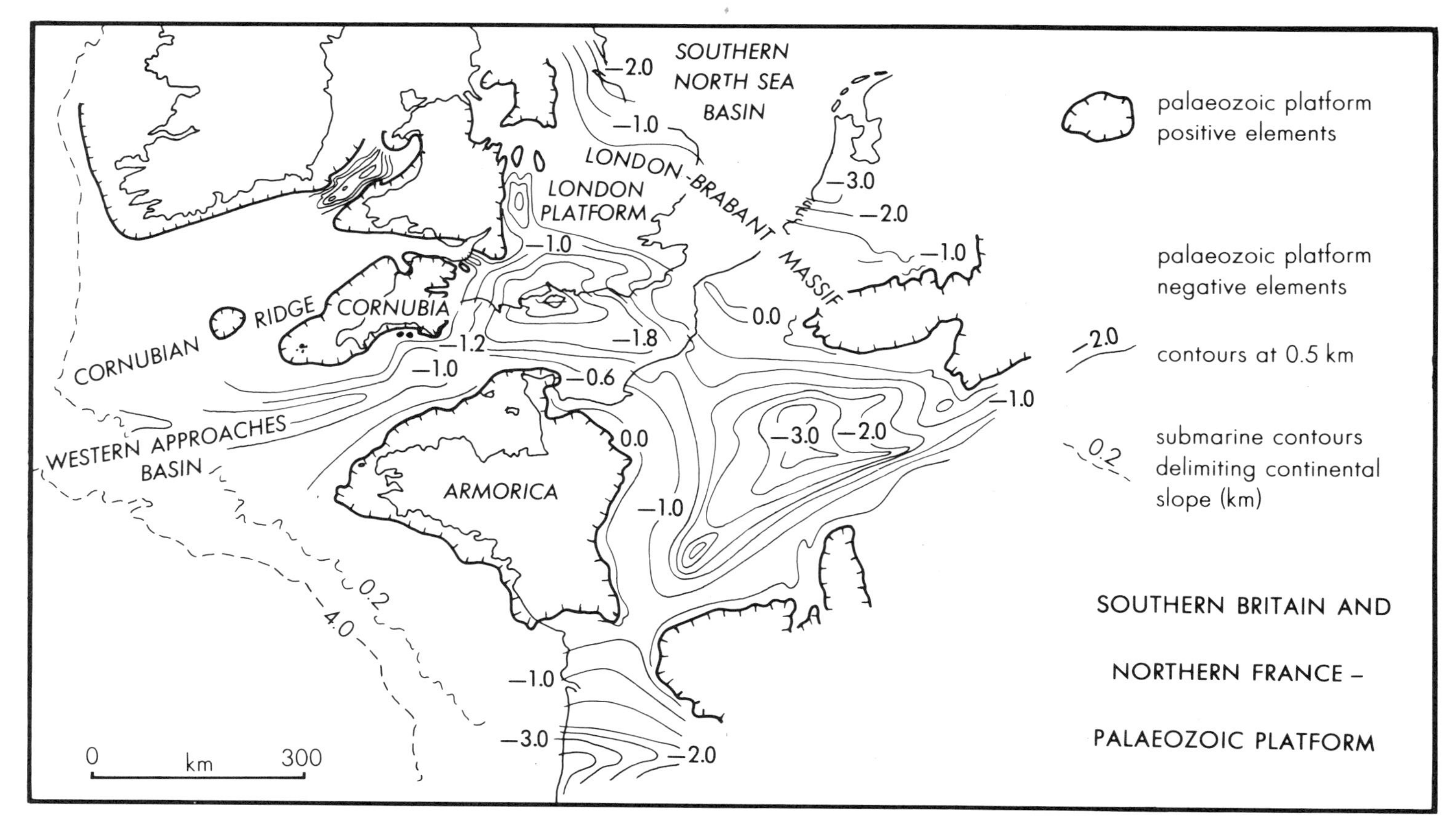


Fig. 6. The Palaeozoic platform of southern Britain and northern France (various sources).

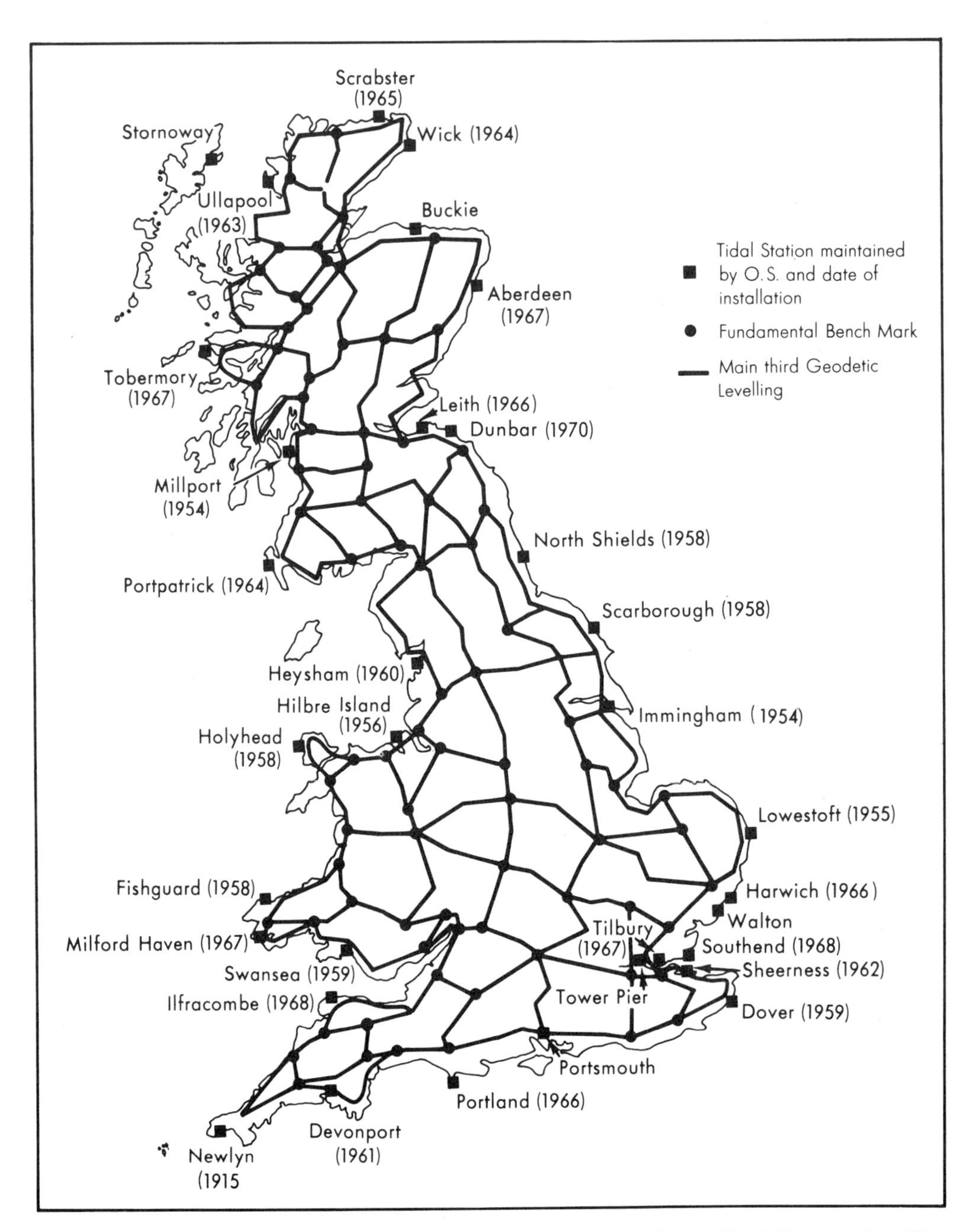


Fig. 7. Tidal stations maintained by the Ordnance Survey and lines of main Third Geodetic Levelling (Eady 1976).

of a surge along the coast recognised (Horner 1977). Low atmospheric pressure raises sea level but more important in the North Sea is the piling up of water by the wind, especially when this coincides with or is close to high tide, the Coriolis effect driving the flow caused by a northerly wind against the east coast. In the English Channel a westerly storm would tend to pile the water against the French coast, an easterly gale being more serious for the English shore.

Law (1975) was one of the first to investigate English Channel storm surges and their coastal effects and George and Thomas (1976) have traced over 100 examples for the 1967–74 period (averaging 14 per year), entering the English Channel from both west and east. The heights above prediction are lower than in the North Sea (maximum 1 m (3.28 ft.)), as might be expected.

Storm surges may cause sudden, intense erosion with long lasting consequences, but archaeologically it is probably the superimposed effect of many surges over centuries that make a coastal site unhabitable, even though mean sea level remains unchanged. Major storms (not necessarily accompanied by surges) have occurred in Europe over the past 1000 years with periods of 13 to 60 years (Brooks 1970). As yet stratigraphical evidence for storm surges has not been clearly defined.

## *Local Influences*

13. Local tidal and other currents—see under (10) above p. 10.

14. Water density variations are caused by temperature and fresh water sources. These may influence the rate and location of peat accummulation and the flocculation (and hence deposition) of river sediments. Stratigraphic sections may therefore vary greatly over short distances, especially near estuaries, in turn causing differential compaction rates.

15. The existence of differential sediment compaction is well known, though there is scope for further quantitative work. Compaction figures of 2% for sands and 80% for peat have been quoted.

16. The above is influenced by man's lowering of the water table by excessive extraction for domestic and industrial use. Differential movements within London may be attributable to this (Longfield 1932) and on a larger scale there is evidence of sea water penetration into aquifers along the east coast.

17. The building of river and sea walls leads to differential drying out of peaty layers, shrinkage and fall of land level.

18. The local coastal and offshore configuration controls the energy of, and long term erosion or deposition by, wind waves and swell. The relationships between offshore morphology, residual tidal currents and coastline change have been carefully established (Robinson 1966) and they change through time, as Robinson has shown.

19. Finally we may mention the construction or demolition of man-made features—breakwaters, moles, groynes—all of which are relevant to individual site studies, usually in their relationship to other local influences.

### *Instrumental Influences*

These apply particularly to studies of contemporary land and sea movements and are listed for the sake of completeness (cf. Wemelsfelder 1971, p. 122).

20. Site of tide gauge.
21. Alteration of site.
22. The accepted base or zero height: 'the best reference level for use in the construction of a graph of mean sea level movements in Europe during the last 15,000 years may be Normaal Amsterdam Peil (NAP) or it may be the centre of gravity of the earth' (Jardine 1976).
23. Unintentional change in zero height.
24. Gauge or benchmark subsidence.
25. Levelling staff orientation during long levelling transverses (see below).
26. Frequency of connection of gauge to land geodetic levelling network.
27. Density deviations in the stilling well.
28. Adjusting and reading of errors.
29. Alterations of instrument or of working methods.

### *Contemporary Relative Sea Level Movements in Britain*

The results of contemporary measurements of sea level changes should be regarded as underpinning the interpretations of those working on past levels. As we have made clear these changes are the resultants of glacial-, tectonic- and geoidal-eustatism, isostatic rebound, differential crustal movements, plus a host of other influences. The consequent relative sea level changes may vary both spatially and temporally.

*Land surveys*. The reality of differential contemporary crustal movements is demonstrated by earthquakes in recent years in the Channel Islands, Sticklepath in Devon and the Leicester–Nottingham area. The known distribution of Pleistocene ice sheets coupled with the concept of the elastic rebound theory suggest that some contemporary crustal warping in Britain should be apparent in a north–south sense. It is, however, just in this direction that two internationally recognised land levelling problems arise:

(i) Because of the north–south ellipticity of the earth, equipotential surfaces converge polewards, rendering difficult the measurements of true orthometric height differences over long distances. This can be overcome by allowing for gravity changes along the levelling line and all levelling data published for Britain are so corrected.

(ii) In alternate positions along a meridional traverse the figured side of the levelling staff faces south and a combination of shadows on the engraved numbers and expansion introduces a systematic error. At 5 microns per station this would amount to 18 cm (7 in.) of error over the 1804 km (1121 miles) between Newlyn (O.S. datum) and John O'Groats (Eady 1976), a figure comparable with published ranges of sea level change.

The three geodetic levelling networks of the Ordnance Survey (carried out in 1840–60; 1912–21;[4] 1951–9) provide a basis for studying land movements. Figures 8a and b compare the three surveys but the official view (Eady 1976) is that 'in all the work carried out so far [by the Ordnance Survey] no crustal movement of sufficient amplitude to be detected by levelling has been found . . . ', an opinion stated earlier by Kelsey (1972) with specific reference to south-east England[5] but qualified thus: 'the comparison of altitudes at fundamental bench marks derived from the two levellings (2nd and 3rd) suggest that, while there are small changes in elevation of the land in southern and central England relative to Newlyn, these are not significant compared with possible errors in the levelling; but in northern England there is an apparent uplift of the land relative to Newlyn of 175 mm (6.75 in.) in the 32 years between the levellings (or approximately 5 mm (.25 in.) per year). This is significant. While land uplifts of this order and greater due to deglaciation have been recorded in northern Scandinavia, this figure for Great Britain cannot be accepted without supporting evidence from some other source such as oceanographic data.' There is no doubt that the earliest survey contained instrumental and datum problems (e.g. Liverpool datum was based on a 10-day period) but for the record it is worth noting that mean sea level at Dunbar appeared to have fallen by several centimetres between both the first and second, and the second and third surveys (Eady 1976), showing consistency, at least, in the direction of change.

Inspection of fig. 8 shows that there is little consistency in pattern of local differential movement to be deduced from the two pairs of surveys, in fact on the basis of these crude figures movements in opposite directions have occurred (cf. the Sudbury area, west of Felixstowe, which apparently sank *c.* 5.0 cm (2.0 in.) between the first pair, rising by the same amount between the second pair). We must not, however, dismiss local contemporary crustal movements. The various reports of the Commission for Recent Crustal Movements (e.g. 1962, 1965) contain ample evidence, based upon national geodetic surveys in various parts of Eurasia, for movements at rates of cm/century.

The time span of tidal gauge records in this country is shown in fig. 9, where it will be seen that Sheerness, Aberdeen and Newlyn have the longest British records for the study of contemporary sea level changes, but the best maintained tidal observatory, that of the Ordnance Survey at Newlyn, has only been operating since 1915. The 1953 North Sea storm surge did stimulate the authorities into setting up 38 tide gauge stations around our coasts (fig. 7) tied to the Ordnance Survey levelling network and checked annually by the Survey. Since 1976 work on them has been coordinated by the Committee on Tidal Gauges. Unfortunately not all the gauges are of equal standard and understandably some sites are more suited to maritime than scientific needs (Lennon 1976). We have, however, come a long way from the attitudes of 1859 when, although observations were made every ten minutes for one complete tide at 24 tidal stations around the British Isles for one complete lunation, Sundays were excepted! (Henrici 1911).

It is generally accepted that, at present, tide gauge records are less accurate, more subject to local disturbance and less securely tied to a national levelling datum than are land surveys. That mean sea level, even when correctly measured, is not a simple surface complicates the interpretation of tide gauge records. Thus Kelsey (1970), discussing the Unified European Levelling Network, concluded

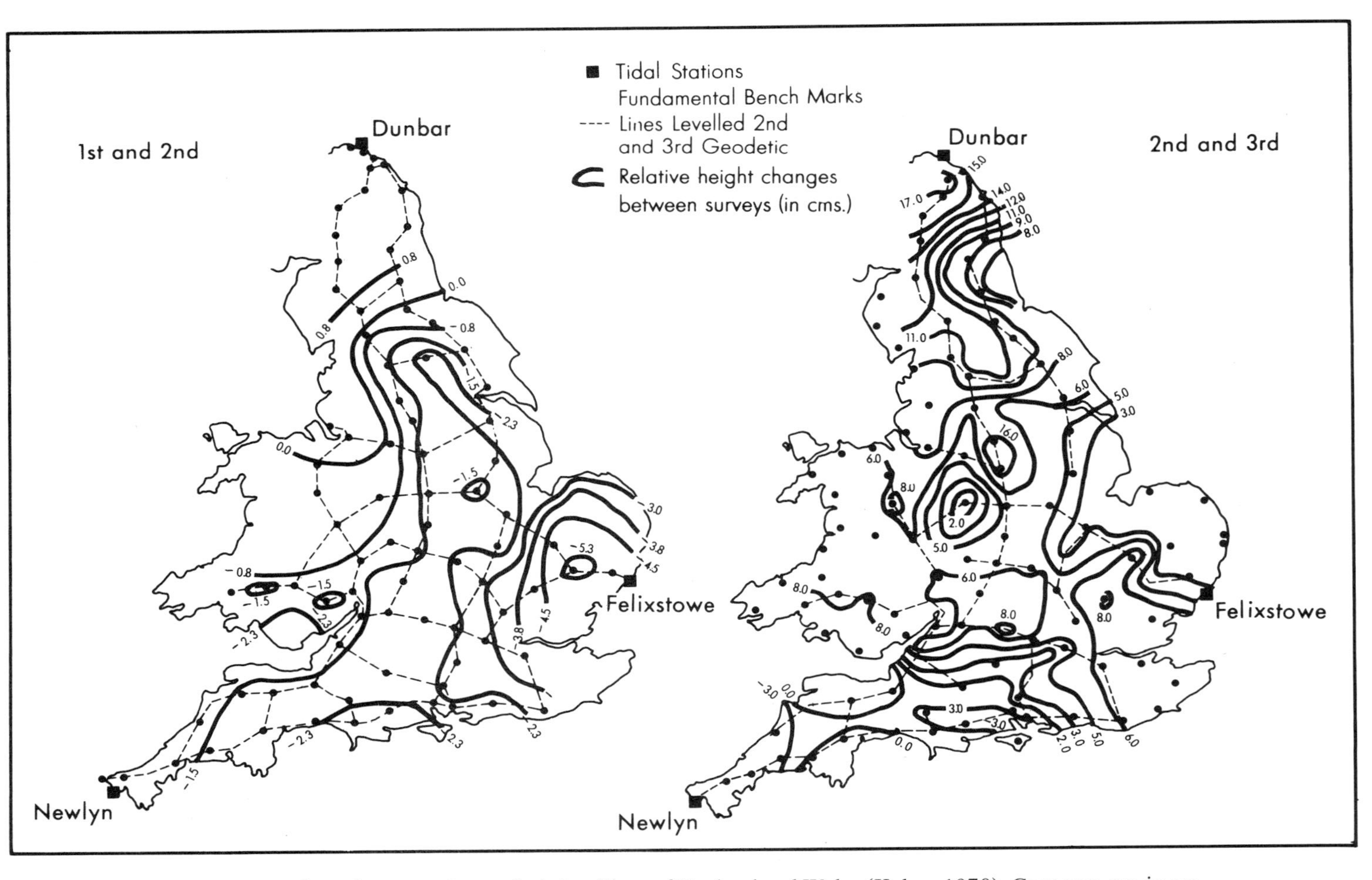


Fig. 8. Comparisons between the geodetic levellings of England and Wales (Kelsey 1970). Contours are in cm.

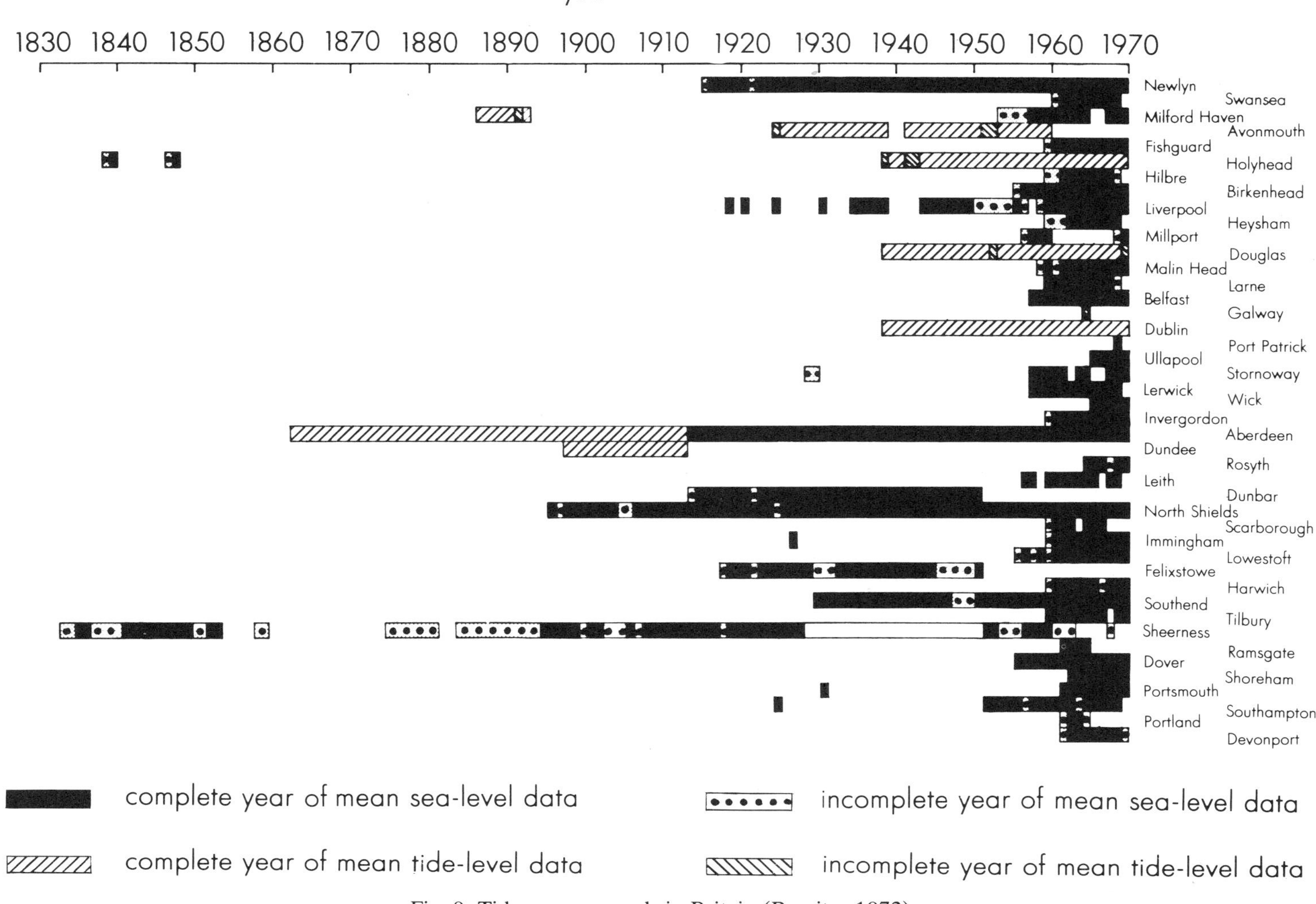


Fig. 9. Tide gauge records in Britain (Rossiter 1972).

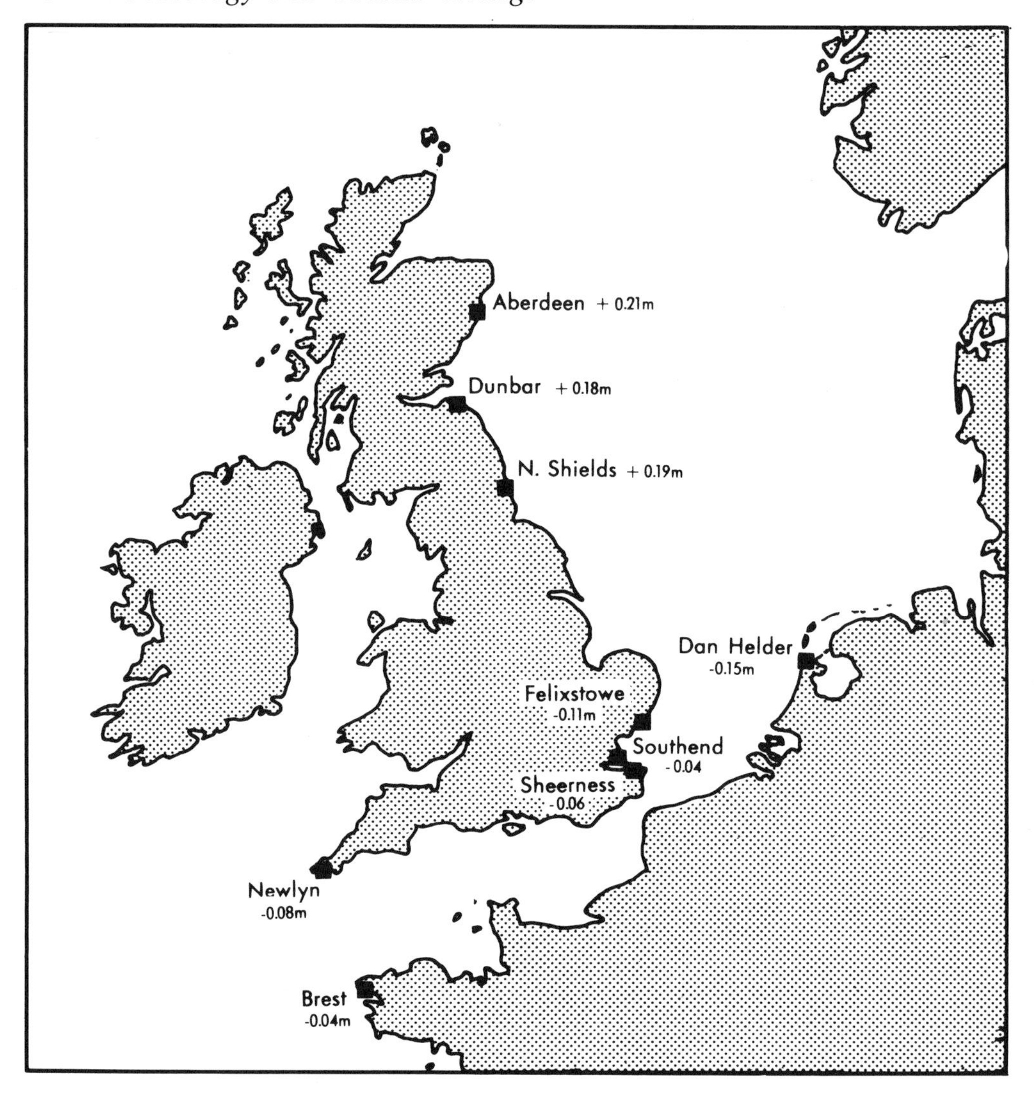


Fig. 10. The topography of the isobaric sea surface around the British Isles corrected to 1970, in terms of the United European Levelling Network (U.E.L.N.) (Kelsey 1970).

that there is a true northward rise in mean sea level along the North Sea (fig. 10), which is unlikely to have been constant in magnitude through time.

This is not the place for a complete review of the often conflicting conclusions of the many studies on mean sea level changes and their causes in Britain but one or two main points can be made.

First, south-east England has not been well served in the past by either geodetic levelling or tide gauging stations. Comparative studies within the area and with other places must be hampered.

Secondly, it must now be apparent that one cannot usefully extrapolate rates of sea level change measured over short periods. Thus even one of the most meticulous workers in this field revised his estimates of relative sea level rise in south-east England from 20 to 30 cm per century (7.75–11.75 in.) in successive reviews (Rossiter 1967, 1972). At the present time (1978) it seems that the rate is smaller.

Thirdly, there is evidence of secular variations of sea level in Britain and these have been set out by Rossiter (1972) as follows (mm/year):

Aberdeen +0.8 Dunbar +0.1

Felixtowe +1.6

Southend +3.4 Sheerness +3.3

Newlyn +2.2 Brest +2.1 (1 mm = .0375 in.)

Fourthly, having regard to the above regional differentiation and as the Ordnance Survey is as yet unable to confirm the presence or absence of contemporary differential crustal movements on the basis of land surveys, attempts have been made to estimate them from the tide gauge records. These records of relative sea level changes are assumed to include a uniform eustatic factor, which, when discounted, should reveal the magnitude and trend of crustal deformation. The pitfalls of such an approach need no elaboration. Rossiter (1967, 1972) endeavoured to allow for as many variables as possible and concluded that, taking as a working hypothesis a eustatic rise of 1 mm (.0375 in.) per year, eastern Scotland is rising at about 0.5 mm per year and southern England is sinking at approximately 1 mm (.0375 in.)/year, reaching over 2 mm (.075 in.) per year in the Thames area (rates of 10 cm (4 in.) and 20 cm (8 in.) per century respectively). Isostatic readjustment is probably a major contribution to these movements.

Fifthly, sea level records can be extended historically to 1791 at London Bridge for since that date high tide has been progressively rising at about 0.73 m (2.5 ft.) per century (Horner 1977), although embanking of the river may have affected the tidal range. Extrapolation to earlier time has been assessed critically by Akeroyd (1972), who estimates that 'the relative positive rise in sea level along the southwest and east coasts over the past 6000 years has been probably at average rates of 10 cm (4 in.) and 13–16 cm (5.25–6.25 in.) per century respectively . . . the average rate will not necessarily be valid for any given area nor for any given period.' Her figures are about one half of the maximum rates that have been quoted for south-east England.

The techniques, instruments and theory used to estimate sea level changes will always be capable of improvement. The form of the geoid, or mean sea level, is a factor often neglected.[6] The effect of local compaction and regional differential movements remains to be quantified. The best approach is undoubtedly the accurate tying in of critical archaeological surfaces to the Ordnance Survey levelling network and the absolute dating of the horizons with organic remains.

## BIBLIOGRAPHY

Akeroyd, A. V. 1966. Changes in relative land and sea levels during the post-glacial in southern Britain. Unpublished M.A. thesis, University of London.

—— 1972. Archaeological and historical evidence for subsidence in southern Britain. *Phil. Trans. R. Soc. Lond.* A. **272,** 151–71.

Anon. 1978. New format for world-wide mean sea level data bank. *N.E.R.C. News Journal,* **2,** 5.
Bott, M. H. P. and Scott, P. 1964. Recent geophysical studies in south-west England, in *Present Views of some Aspects of the Geology of Cornwall and Devon*, Hosking, K. F. and Shrimpton, G. J. (eds.). Royal Geological Society of Cornwall, Penzance.
Brooks, C. E. P. 1970. *Climate through the Ages* (2nd edn.). New York.
Brooks, M. 1973. Some aspects of the Palaeozine evolution of western Britain in the context of an underlying mantle hot spot. *Journ. Geol.,* **18,** 81–8.
Brown, E. H. 1960. The building of southern Britain. *Z. Geomorph.* **4,** 264–74.
Cartwright, D. E. and Crease, J. 1963. A comparison of the geodetic reference levels of England and France by means of the sea surface. *Proc. Roy. Soc. Lond.* A. **273,** 558–80.
Churchill, D. M. 1965. The displacement of deposits formed at sea level, 6500 years ago in southern Britain. *Quaternaria,* **7,** 239–47.
(C.R.C.M. 1962, 1965, etc.) Int. Union geodesy and geophysics: *Commission for Recent Crustal Movements.*
Currie, R. G. 1976. The spectrum of sea level from 4 to 40 years. *Geophys. Journ. Roy. Astron. Soc.,* **46,** 513–20.
d'Hieres, G. C. and Le Provost, C. 1970. Spatial variations of the mean sea level in the English Channel. *Int. Union geodesy and geophysics, report on Symposium on coastal geodesy.* Munich, 427–37.
Delany, F. M. (ed.) 1971. *The geology of the East Atlantic continental margin.* Reps. 70/13–15. Inst. Geol. Sci., London.
Denton, G. H. and Karlen, W. 1973. Holocene climatic variations—their patterns and possible course. *Quaternary Res.,* **3,** 155–205.
Eady, J. 1976. The monitoring of tidal gauges by the Ordnance Survey. *Proc. Symposium on tidal recording, Southampton, Hydrogr. Soc. Sp. Publ.* No. 4, 37–45.
Everard, C. E. in press. Marine geomorphology. Mean sea level. *Progress in Physical Geography.* London
Fairbridge, R. W. and Krebs, O. A. 1962. Sea level and the southern oscillation. *Geophys. Journ. Roy. Astron. Soc.,* **6,** 523–45.
Farrell, W. E. and Clark, J. A. 1976. On postglacial sea level. *Geophys. Journ. Roy. Astron. Soc.,* **45,** 647–67.
Flint, R. F. 1971. *Glacial and Quaternary Geology.* New York.
George, K. J. and Thomas, D. K. 1976. Two notable storm surges of the south coast of England. *Hydrogr. Journ.,* **2,** 13–16.
Geyhl, M. A. and Streif, H. 1970. Studies on coastal movements and sea-level changes by means of the statistical evaluation of C14 data. *Int. Union geodesy and geophysics, report on Symposium on coastal geodesy.* Munich, 599–611.
Gordon, D. L. and Suthons, C. T. 1963. Mean sea level in the British Isles. *Admiralty Science Publication* No. 7, Hydrographic Department, London.
Henrici, E. O. 1911. Mean sea level. *Geog. Journ.* **39,** 605–6.
Hoinkes, A. 1960. Neue Ergebnisse der glaziologischen Erforschung der Antarktis. *Umschau,* **19,** 627–30.
Horner, R. W. 1977. *Conference on Thames Barrier Design.* Institution of Civil Engineers (mimeograph).
Jardine, W. G. 1976. Some problems in plotting the mean surface level of the North Sea and the Irish Sea during the last 15,000 years. *Geol. Foreningens Förd.,* **98,** 78–82.
Johnson, D. 1929. Studies of mean sea level. *Bulletin of the National Research Council.* No. 70. National Academy of Science, Washington D.C.
Kelsey, J. 1970. Considerations arising from the 1970 readjustment of the geodetic levellings of Great Britain. *Int. Union geodesy and geophysics, report on Symposium on coastal geodesy.* Munich, 331–7.
—— 1972. Geodetic aspects concerning possible subsidence in south-eastern England. *Phil. Trans. R. Soc. Lond.* A. **272,** 141–9.
King, L. C. 1950. The study of the world's plainlands. *Quart. Journ. Geol. Soc. Lond.,* **106,** 101–31.
King, W. B. R. 1954. The geological history of the English Channel. *Quart. Journ. Geol. Soc. Lond.,* **110,** 77–98.
Lamb, H. H. 1972. British Isles weather types and a register of the daily sequence of weather patterns, 1861–1971. *Met. Off. Geophys. Mem.* No. 116, H.M.S.O.

Law, C. R. 1975. Storm surges in the English Channel. *Hydrogr. Journ.* **2,** 30–34.
Lennon, G. W. 1976. The national network for the monitoring of sea-level and the Committee on Tide Gauges. *Proc. symposium on tidal recording, Southampton, Hydrogr. Soc. Sp. Publ.* No. 4, 5–11.
Lisitzin, E. 1974. *Sea Level Changes.* Amsterdam.
Longfield, T. E. 1932. The subsidence of London. *Ordnance Survey Professional Paper* No. 14. London, H.M.S.O.
Mitchell, G. F. 1977. Raised beaches and sea levels: in *British Quaternary Studies: Recent Advances,* F. E. Shotton (ed.).
Mörner, N. A. 1976. Eustasy and geoid changes. *Journ. Geol.,* **84,** 123–51.
—— 1977. Geoidal eustasy: a new factor in sea level studies as well as in palaeoclimatology and earth's geophysics. *Abs. Woods Hole Oc. Inst. Symp.,* No. 2.
Munk, W. H. and MacDonald, G. F. 1960. *The Rotation of the Earth.* London.
Paterson, W. S. B. 1972. Laurentide ice sheet: estimated volumes during Late Wisconsin. *Rev. Geophys. Space Phys.,* **10,** 885–917.
Perry, A. H. and Walker, J. M. 1977. *The Ocean-Atmospheric System.*
Pirazzoli, P. 1976. Les variations du niveau marin depuis 200 ans. *Memoires, Laboratoire de Geomorphologie, l'Ecole Practique des Hautes Etudes, Dinard.*
Robinson, A. H. W. 1966. Residual currents in relation to shoreline evolution of the East Anglian coast. *Marine Geol.,* **4,** 57–84.
Rossiter, J. R. 1967. Annual sea level variations in European waters. *Geophys. Journ. Roy. Astron. Soc.,* **12,** 259–99.
—— 1972. Sea level observations and their secular variation. *Phil. Trans. R. Soc. Lond.* A. **272,** 131–7.
Ters, M. 1973. Les variations du niveau marin depuis 10,000 ans le long du littoral atlantique français. *Assoc. Français pour l'Etude de Quaternaire, Suppl. au Bull.* no. 36, 114–42.
Tooley, M. J. 1976. The I.G.C.P. project on sea-level movements during the past 15,000 years. *Quaternary Newsletter* No. 18, 11–12.
Turekian, K. K. 1976. *Oceans* (2nd edn.). New Jersey.
Umbgrove, J. H. F. 1947. *The Pulse of the Earth* (2nd edn.). The Hague.
Walcott, R. I. 1972. Past sea levels, eustasy and deformation of the earth. *Quaternary Res.,* **2,** 1–14.
—— 1975. Recent and late Quaternary changes in water level. *Trans. Am. Geophys. Union,* **56,** 62–72.
Wemelsfelder, P. J. 1971. Mean sea level as a fact and as an illusion. *Inst. Hydr. Rev.,* **48,** 115–27.
West, R. G. 1968. *Pleistocene Geology and Biology.* London.
—— 1972. Relative land-sea level changes in southeastern England during the Pleistocene. *Phil. Trans. R. Soc. Lond.* A. **272,** 87–99.

## REFERENCES

[1] See West, 1968, p. 224 for a discussion of stratigraphical terminology.
[2] It has recently been claimed that at about 5000 bp the sea around the British Isles was *c.* 4 m (13.11 ft.) above its present level (Mitchell 1977).
[3] Revised to 0.667 m (*c.* 2 ft.) by Kelsey (1970).
[4] Extended to Scotland in 1936–52.
[5] Unfortunately the second geodetic levelling did not include south-east England or East Anglia.
[6] See Permanent Service for Mean Sea Level (Anon. 1978).

# Caerleon and the Gwent Levels in Early Historic Times

G. C. Boon, F.S.A.

The legionary fortress of Caerleon lies on a terrace above a wide-swinging loop of the tidal Usk, 14 km (8.5 miles) upstream of its confluence with the Severn. To the south-west, there is a broad expanse of alluvium, where excavations in 1963 revealed a stone-built Roman quay, lying a good 230 m (750 ft.) within the line of the present river-bank.[1] The quay had been built in the early third century, and extended in the late third century: the discovery therefore identified the position of the right bank of the Usk at a particular spot, level, and time within a flood-plain of grey-blue estuarine alluvium here some 500 m (1,600 ft.) wide. The height of the metalling behind the quay, and its date, throw useful light on the character of Romano-British settlement in the Gwent Levels and the opposite Levels in the County of Avon; and the post-Roman history of the quay-site leads into a consideration of the state of the Levels in early medieval times, and of the first attempts to win agricultural land by drainage and enclosure in South Wales.

*The Roman Period*

The height of the hard-standings behind the Caerleon quay was, upon a close average, 6.56 m (21.52 ft.) above Ordnance Datum. The significance of this figure will appear below; suffice it to add here that tidal levels at Caerleon are only 10 cm (4 in.) higher than at the mouth of the Usk, so that the alluvial surface, at any given period, would on the whole lie at a very similar height both at Caerleon and in the Levels bordering the Severn. Now, it was for many years accepted that the later Roman period saw a disastrous marine transgression, when these lands were buried beneath as much as 3 m (10 ft.) of alluvium. The suggestion was Sir Harry Godwin's, and was based on the undoubted occurrence, at various spots in the

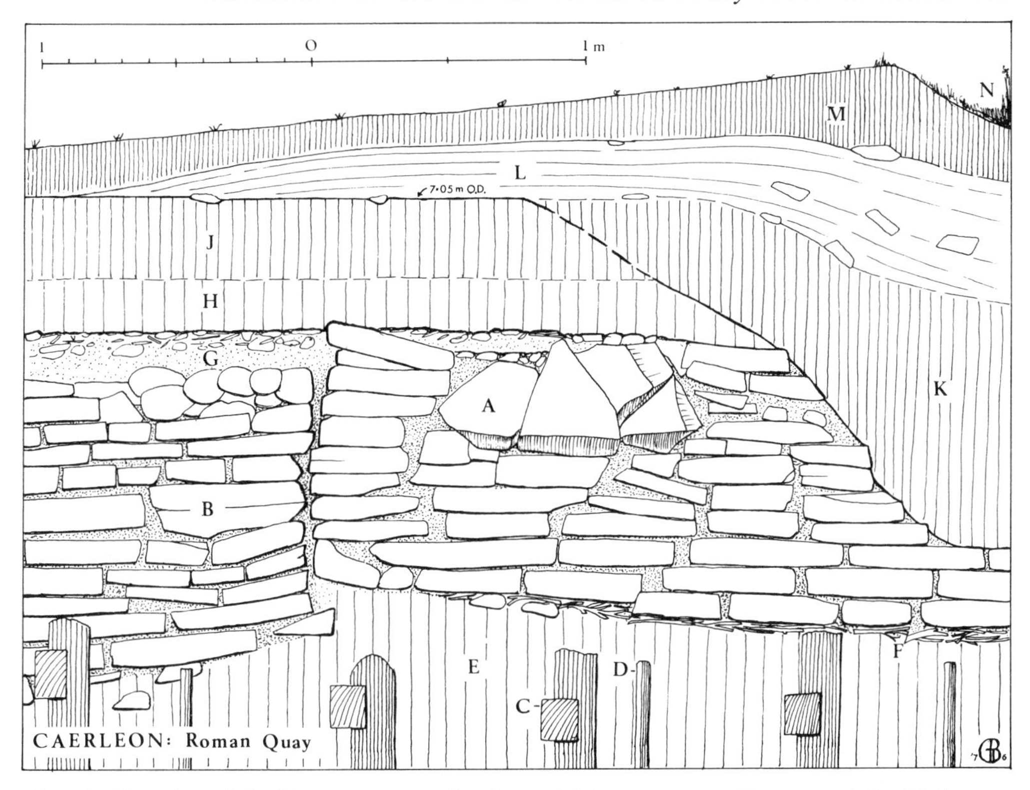


Fig. 11. Elevation of the Roman quay at Caerleon, third century A.D. Note especially: H, J—grey and brown alluvium over the Roman metalling; K—ditch on line of present hedgerow.

Somerset Levels, of Roman pottery lying at such a depth.[2] In 1971 a paper by Stephen Locke assembled the evidence from the Caldicot Level of Gwent and reached a similar conclusion, although it was prudently pointed out that none of the Roman remains hitherto recovered from the area could be regarded as stratified.[3] The height of the third-century quay at Caerleon, which lay within 50 or 60 cm (20–24 in.) of grass (fig. 11), introduced a novel and discordant note.

A new point of departure, however, was offered by A. B. Hawkins, also in 1971.[4] He drew upon discoveries by the North Somerset Archaeological Research Group in the Level near Clevedon (fig. 13),[5] which showed that the occupied surface of the first to fourth centuries was only 20–60 cm (8–24 in.) below that of today in an area of some two dozen square km. Datable objects ranged from late Iron Age pottery to a coin of Magnentius (A.D. 350–3). Hawkins's solution to the paradox which thus confronted him in the light of Godwin's records of deep-lying material was that this latter had been merely refuse, cast into creeks and inlets which dissected a higher habitable surface. At first sight, this explanation seems unlikely or even feeble; but if we look at the state of the coast beyond the sea-wall at Rumney Great Warth, east of Cardiff (figs. 12,13), it takes on a good deal of colour. Here, the mud-cliff at the edge of the flat alluvial belt is penetrated by vertical-sided inlets. Such inlets must have existed in Roman times when the coast was subject to erosion, as it is today, and they could well have received rubbish from settlements at the higher level. It may be added that they *demonstrably* received shards washed out from their original positions within the mud-cliff; and upon occasion the fury of the waves has tossed such material upon the greensward, where I have found it, as well as in the bottoms of the inlets. The mechanics of erosion and redeposition of datable material below, and even above, its original stratigraphical horizon suggest a slight modification of Hawkins's simple reading of the evidence.

At one spot at Rumney Great Warth, a little pottery was found in its original position—the first time ever recorded along the Gwent coast. It lay between 6.44 and 6.64 m (21.13 and 21.78 ft.) above O.D., and thus answered to the 6.56 m (21.52 ft.) of the third-century hard-standings behind Caerleon quay, as well as to the discoveries on the Somerset side. There is a good deal of material from the Rumney site, which has been almost wholly eroded; among it are shards of late Iron Age to third-or fourth-century date, and two or three coins, of which the latest was struck about A.D. 330. Elsewhere along the Gwent coast late Iron Age and Roman pottery has been found on the foreshore, at about half-a-dozen spots. Of these the best-known is at Magor[6] where a fairly large settlement seems to have been almost totally eroded. 400 m (1,300 ft.) inland a new sewage-works was built in 1966, and at the base of a deep shaft a small group of unabraded pottery of very mixed Roman date was recovered at about 3 m (10 ft.) above O.D.,[7] the same depth, approximately, as the beds of the inlets at Rumney Great Warth, and as the foreshore at Magor itself. The most probable explanation of this material is that it had accumulated in the bottom of a long inlet, or watercourse, of which no surface sign now remains. Such a buried inlet has been traced at Highbridge in Somerset by S. G. Nash.[8]

It was Hawkins's view that the condition of the Levels was in Roman times intermediate between mud-flat and salt-marsh, and attracted only seasonal frequentation. That may well have been so, in parts; but elsewhere the evidence leads to a more positive and striking conclusion. There can be no doubt at all but that large tracts were permanently settled. On some of the sites near Clevedon, signs of permanent buildings were found, and remains of a corn-drying plant can but imply that some of the ground had been suitable for arable farming, which in its turn implies drainage.[9] Even more revealing is the testimony of an important, but neglected villa at Wemberham, near Yatton (fig. 13), not far from some of the sites discovered by the North Somerset Archaeological Research Group.[10] The remains, found in 1884 during land-drainage, lie about 80 cm (30 in.) below grass. The excavations suffered from all the shortcomings of work at that period, but we see that the site had been occupied from the second century until the fourth, when the house was thoroughly rebuilt to incorporate a sizeable bath-suite and a reception-room, floored in elegant mosaics (fig. 14), nearly 10 m (33 ft.) long. It need scarcely be pointed out that such a low-lying spot—traversed today by a little embanked river—would not have been chosen for the residence of a family of good standing, were the surroundings not generally dry and pleasant, and the yield of the estate sufficient to maintain the household in comfort.

It is not practicable in this paper to allude to many other sites; but one which further illustrates the exploitation of the alluvium is Kingsweston villa near Avonmouth.[11] It lies at 7.5 m (25 ft.), within a stone's-throw of the landward edge of the Severn flood-plain, one of a number of settlements – the others subordinate,

Fig. 12. The coast at Rumney Great Warth, 1975, looking towards Cardiff (photo: G.C.B.).

most probably—in a narrow band extending for some 3 km between 9 and 12 m (30–40 ft.) above O.D. (fig. 13). There can be little doubt but that these buildings were placed where they were in order to farm the Level to the north. Kingsweston was first occupied, like Wemberham, in the second century A.D., and like it, too, was completely rebuilt towards 300. It was surely the prospect of a rich reward from these lands that directed new money to both places—at Kingsweston, perhaps foreign money, if Branigan's theory regarding detailed affinities of plan between the enclosed-courtyard design at Kingsweston and among the villas of a ravaged northern Gaul and Rhineland is as correct as it is attractive.[12] As we saw, however, the latest coin from the Clevedon Level is of *c.* A.D. 350; indeed, the latest from Wemberham itself is of *c.* 330, like the latest from Rumney Great Warth on the opposite shore; and it would seem that investment came too late. At Kingsweston, the cultivation of the slopes south of the villa maintained occupation on a reduced basis until the third quarter of the century, and there were squatters later still; but the middle years of the fourth century were, beyond much doubt, the time when Roman farming on the Levels met disaster 'by reason of the outrageous flowing, surges and course of the sea in and upon marsh-grounds and in other low places heretofore by politick wisdom won and made profitable,' to quote from the preamble of a statute of 1531 setting up Commissions of Sewers.

The most celebrated object from the Severn Levels is the inscribed stone from Goldcliff, Gwent (fig. 13). It was discovered in 1878 about 3 m (10 ft.) above O.D., projecting from the eroded mud-cliff well ahead of the present sea-embankment with which, through an error of Prebendary H. M. Scarth's, it unhappily remains indelibly connected.[13] The inscription, towards one end of the

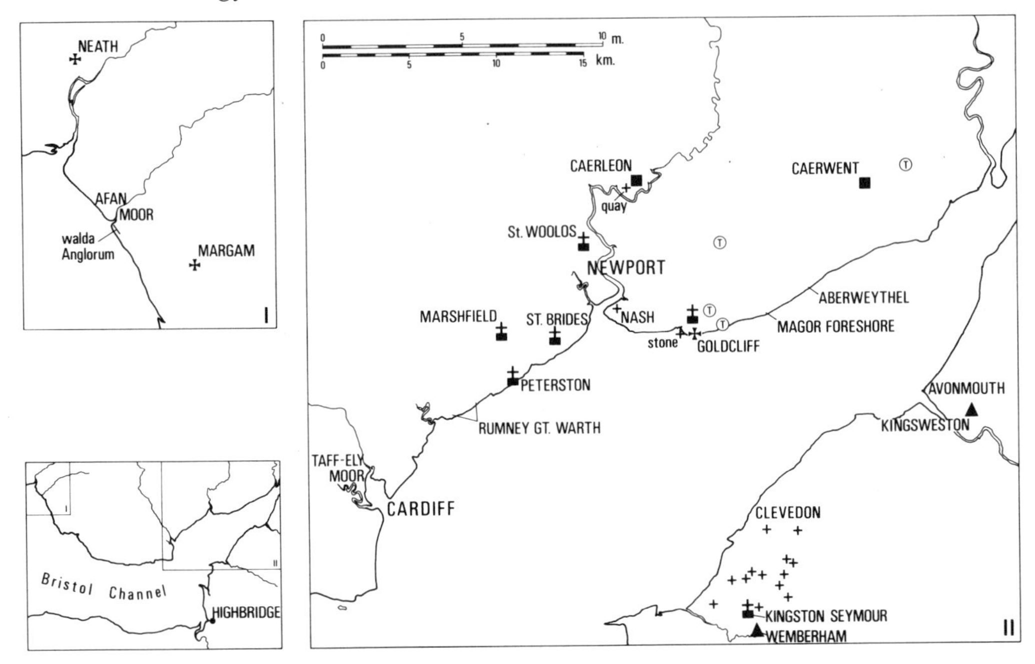


Fig. 13. The Severn Levels: sites mentioned in the text. Note: small crosses near Clevedon—sites discovered by the North Somerset Archaeological Research Group; T in circle—*ton* names (see footnote 36).

slab, is to the effect that the Century of Statorius Maximus, of the first cohort—naturally, of Second Augustan Legion at Caerleon about 9 km (5.5 miles) to the north—constructed 33½ paces of some unspecified work.[14] R. P. Wright's reading of the numeral, and his deduction that a length of 100 paces was divided into thirds for different gangs, allow us to believe that the work was of considerable magnitude to have been marked in this way. Goldcliff, it may be explained, is a low and much-eroded knoll, quite isolated and therefore prominent in that flat countryside; the coincidence of the findspot suggests that the linear work commemorated may have served as a boundary between the legionary lands and those of the neighbouring Silurian *civitas*, whose capital at Caerwent is only 13 km (8.0 miles) distant. Secondly, because the stone is so shaped as to have been plainly intended to be set in or beside an earthwork, we may infer that this was either an embanked natural watercourse, similar to the present stream which runs across the Level from the high ground to the north, and debouches upon the Severn as Goldcliff Pill; or an artificial drain. The inscription was of the roughest, and it is therefore difficult to date; but it can hardly be early Roman, and is most probably of the later second or the third century A.D. The legion left Caerleon in the 290s.[15]

The stone declares a military interest in the Level; and there can be little doubt in what that interest lay. The Gwent Levels have offered superb resources to the pastoralist. In 1815 the agronomist Charles Hassall wrote of the excellent grasses, of dairying, of summering of cattle, and of raising colts;[16] at an earlier date, too, the warths bordering the Severn were common-pasture to which each parish

Fig. 14. Mosaic pavement from the inner part of the Reception Room, Wemberham villa.

extended, or were in part made the subject of valuable benefactions to religious houses. We can guess, therefore, the value placed on this tract of southern Gwent by the legionary *pequarii*, or stockmen, and by the civilian occupants. Of pastoral farming we have indeed much evidence from animal-bone, sometimes found in fair amount mixed with Roman pottery: it is chiefly of ox, but sheep are also well represented.[17] To define the legionary *territorium* would have been necessary from the first, so that the earthwork discussed will have had some predecessor; to improve it by drainage, or to protect it from the overflow of waters, will have been the next step.

## *The Post-Roman Period*

The profile above the hard-standings at Caerleon quay (fig. 11), which was eventually buried by about 50 cm (20 in.) of alluvium (bringing the surface to 7.05

m (22.97 ft.) above O.D., or 10 cm (4.0 in.) below grass, offers a key to the post-Roman history of the Levels. Unfortunately, the departure of the legion in the 290s, and the rarity of fourth-century material outside the fortress itself, render valueless the late third-century *terminus post quem* from the quay for the onset of permanent inundation: the late Constantinian date applying to the abandonment of farming on the Levels is a better guide. The section above the metalling showed an abrupt colour-change from grey to brown, and this was at first taken to indicate that the first 15 cm (6.0 in.) or so of grey clay was of estuarine origin, and the upper brown clay, of riverine; but in point of fact the two layers could not be cleanly detached, and both were shown to be intertidal by Dr. John Haynes of the University College of Wales, Aberystwyth, who studied the foraminifera which they contained. He also reported abundant flecks of ash, plant-tissues and seed-cases in the grey clay, from about the surface of which a few shards of thirteenth- to early fourteenth-century pottery had been recovered. The evidence suggests that a phase of equilibrium in the onset of floods and silt-deposition had occurred by about 1300, when the land could again be frequented and put to use. The *inquisitio post mortem* of Gilbert de Clare, 1314, makes mention of pastures called 'the warth' and 'the small warth' at Caerleon.[18]

Such an interval corresponds with earlier pauses archaeologically detected at the same spot;[19] and its duration is suggested by the presence of only sixteenth- to nineteenth-century pottery in the weathered, brown upper clay. The stable conditions of earlier medieval times seem to have been followed, until the sixteenth century, by severe inundation and alluvial deposition; and at such a period of rising sea-level, the effect of storm-floods will have been grievous. The first statute governing Commissions of Sewers is not that of 1531, quoted above, but that of 1427;[20] before, there were local and non-statutory commissions. Serious floods include those of 1484[21] and 1606–7,[22] and the latter is reported to have caused havoc from Carmarthenshire up to Gloucester, and far up the tributaries. Commemorative plaques in Goldcliff and Peterston Wentloog churches, 10 km (6 miles) apart on the Gwent Levels, show that it touched 7.14 m (23.4 ft.) above O.D.,[23] practically identical with the top of the alluvium at Caerleon which, thus, was very probably then deposited. The wide and deep drain cut through the Roman quay-wall (fig. 11, right) could well mark the reclamation of the area afterwards: its line is that of the present ditch and hedgerow, and this in turn corresponds with the limit of embanked and drained fields shown in an estate-plan of 1765 (fig. 15). It is interesting to see that even at this date the common-land, within as well as beyond the Morgan estate, remained unembanked.

The plaque in Goldcliff church (fig. 16) demonstrates the occasional hazard of life on the Levels before the completion of the system of sea-banks; it is dated 1609, and we can well imagine that three years did indeed elapse before recovery was complete. The text is, after a fashion, metrical:

> On the twenty day of Janu'ry e'en as it came to pas
> It pleasèd God the flud did flow to th'edge of this same bras;
> And in this parish theare was lost five thousand and od powndes
> Besides two and twenty people was in this parrish drown'd.

The events of that January day, however, are curiously refigured in the Life of St. Gwynllyw, written about 1130.[24] The compiler had at his disposal a vernacular

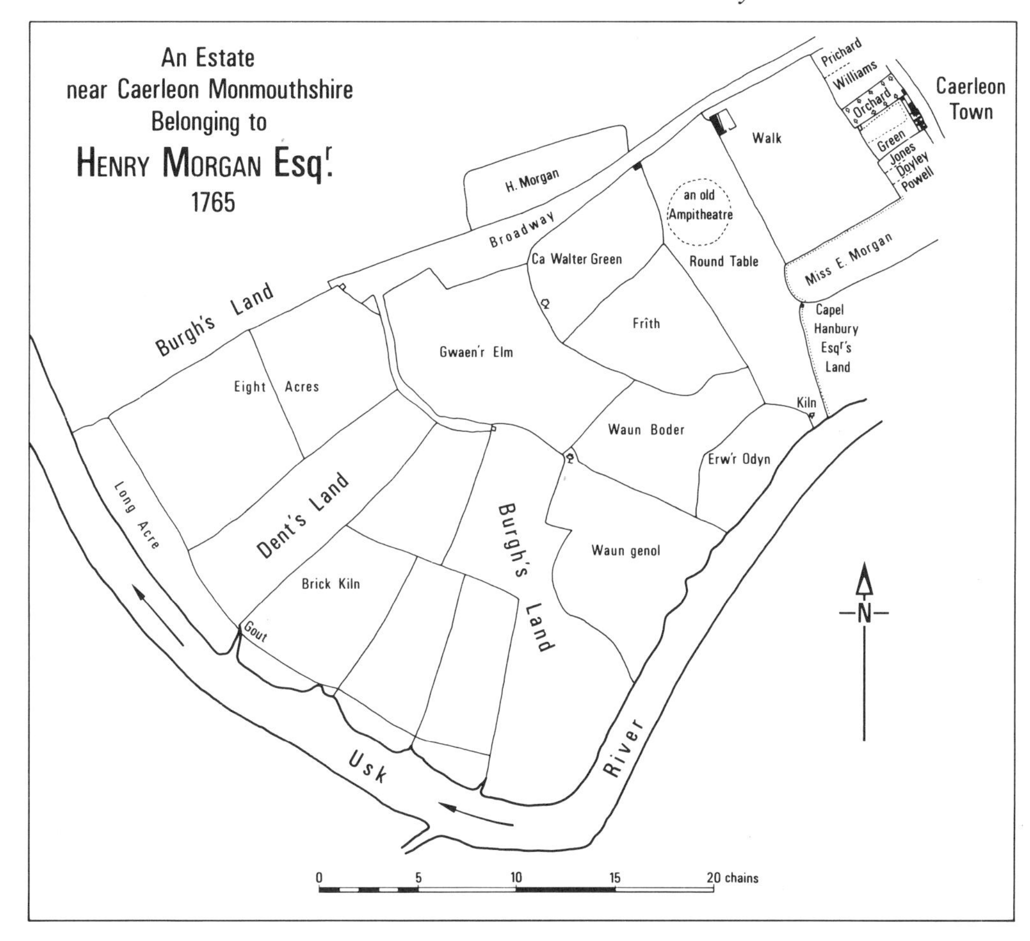


Fig. 15. From a photostat in the National Library of Wales. Note: unembanked 'Burgh's Land', and the use of the alluvium for brick- and tile-making ('Kiln', 'Brick Kiln', 'Erw'r Odyn' = 'Kiln Acre'). The site of the Roman quay is in 'Burgh's Land' immediately opposite the end of the drain dividing 'Eight Acres'. The lateral hedge line is that noted in fig. 11.

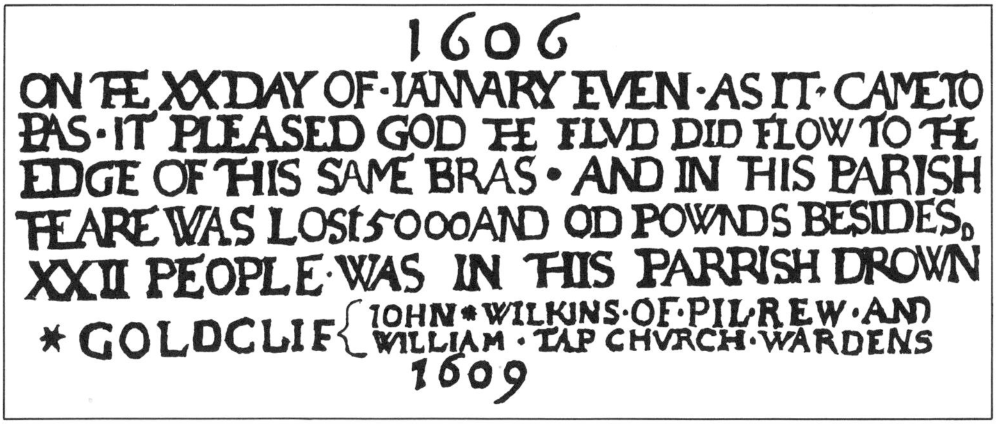

1606
ON FE XX DAY OF·IANVARY EVEN·AS IT·CAME TO
PAS·IT PLEASED GOD FE FLVD DID FLOW TO FE
EDGE OF THIS SAME BRAS·AND IN THIS PARISH
FEARE WAS LOST 5000 AND OD POWNDS BESIDES,
XXII PEOPLE·WAS IN THIS PARRISH DROWN
* GOLDCLIF { IOHN * WILKINS·OF·PIL·REW·AND
WILLIAM·TAP CHVRCH·WARDENS }
1609

Fig. 16. From a rubbing of a brass in the chancel of Goldcliff church (20 × 8 cm).

poem in honour of the saint, of course with a strong miraculous content; other miraculous doings along the Gwent coast recorded in the *Life* suggest that the material dated from shortly before the Norman conquest.

*Britannus quidam versificator, Britannicè versificans*, had (we are told) composed three of a projected four cantos of praises, and was stuck on the fourth, when a violent flood arose. Between the Severn and the church of St Gwynllyw—now St Woolos' Cathedral, Newport, prudently set on a hill overlooking the Levels—no living person escaped, except the poet. When the flood still remained within the confines of the Severn—*manens inter maritima Saverne*—inspiration at last struck our poet; and while his house steadily filled with water he climbed first on to the roof-beams and then on to the roof itself, pouring out his lay. The saint was charmed; and when the flood went down the poet found that his house alone still stood. Details, such as cattle and horses swimming in the flood-waters, add to a picture of a rich pastoral countryside, well-populated, which on our reading of the evidence for Roman times we have little reason to reject.

Among details paralleled in the accounts of the 1606–7 disaster, the reference to the flood being seen while it was still confined by the banks of the estuary is the most curious. In the London newsletters recording eye-witness reports of the disaster, we read, for example, ' . . . about nine in the morning, the sunne being fayrely and brightly spred, huge and mighty hills of water were seen, tumbling over one another . . . to the inexpressible astonishment and terror of the spectators, who at first mistaking it for a great mist or fog, did not on the sudden prepare to make their escape . . . but on its nearer approach, which came on with great swiftness, they perceived it was the violence of the raging seas which seemed to have broken their bounds. . . . In less than five hours' space, most part of the countries on the Severn's banks were laid under water, and many hundreds of men, women and children perished . . . ' Camden, who was revising his *Britannia*, collected the classic facts:[25] a combination of spring tides driven up a narrowing estuary by a three-day gale, reinforced by a stronger blast. It says much for the gradual improvement of the sea-defences along our coast that the highest tidal flood recorded in recent years at Caerleon was well over the 1606–7 mark at 7.95 m (26.08 ft.), and little hurt was suffered.[26] The thoroughly piecemeal origin of the sea-banks along the Gwent coast is well exemplified by two albums of highly accurate and beautifully-drawn and coloured surveys, executed in 1830–1, in the offices of the Caldicot and Wentlooge Levels Drainage Board at Newport. Sheet by sheet, they bring out the local and individual character of the embankments, which ultimately spring from the early reclamation-work to which we now turn. In several parts no banks at all are shown, as we saw in miniature on the Caerleon estate-plan; and sometimes schemes of later date, long since implemented, have been pencilled in.

### *Early Medieval Drainage and Embankment in Glamorgan and Gwent*

In his admirable book, *The Draining of the Somerset Levels*, Michael Williams says that no drainage-works are known to have been carried out before the late twelfth century, although there had been embanking along the river Axe by 1129.[27] South Wales is unlikely to have seen embankments at a much earlier date: we can certainly dismiss as an anachronism, for example, the tale in the *Life of St. Illtyd*, in which the saint thrice and vainly constructed 'an immense dyke of mud and stones' across the mouth of the valley where his *clas* lay.[28] This touch of local colour was probably inspired by early works along the South Wales coast, into which we may now enquire. The early Middle Ages were a period of accretion, as is shown by references to *novus mariscus*, 'new marsh', and the like. One of especial interest is an agreement of 1126 between Urban, Bishop of Llandaff, and Count Robert of Gloucester concerning 100 acres (40.5 ha) in the marsh between the rivers Taff and Ely, at Cardiff.[29] This area could be used, it is stated, 'for ploughing or pasture'—*ad arandum vel ad pratum*. There is no reference to works; but the possibility of arable cultivation is of great significance because, to get the soil in proper heart for the plough, it would have been necessary to drain it, and so bring the salt-content of the soil-water down to well under ½ per cent.; even then, ten years would probably have to be allowed before the land was really ready for the plough.[30]

The impetus to use and indeed to develop coastal alluvium was strengthened in South Wales, as elsewhere, by the foundation of religious houses close to the sea: the alien priory of Goldcliff in 1113, and the Cistercian abbeys of Neath and Margam in 1130 and 1147.[31] Charters mention embankments with some frequency, under the name of *walla, wallum* or *walda*; and of these the earliest is the *walla* or *walda Anglorum* in Afan Marsh near the present town of Port Talbot. The first references to this 'Englishmen's Wall' are of the late twelfth to early thirteenth centuries;[32] and the name shows that workmen with special skills were introduced by the Margam monks soon after the abbey itself was founded. In a phase of accretion, the formation of warths could be expedited by casting brushwood on the mud-flats in order to retain as much tidal silt as possible, so building up the level; and in this way, or perhaps naturally, the growth of land outside the *walda Anglorum* gave the monks a *novus mariscus* which was itself soon protected by a *nova walda*; and when they were given rights over this new land, these were expressed in terms which envisaged cultivation.[33] In a similar sequence on the Taff-Ely moor at Cardiff, we find from a charter of 1226 that Margam was granted 'all our moor which lies outside the new *walda* of Cardiff . . . between the *walda* and the sea . . . besides ten acres of land within our new *walda* to build themselves a sheepfold.'[34] Margam's holding in the region is commemorated by the name of an inner suburb of Cardiff, Grangetown, where an old farmhouse survives to mark the site of the buildings of this grange.

In the fourteenth and fifteenth centuries the pendulum swung back, and erosion gained the upper hand. Goldcliff Priory is shown as a prosperous place in the *Taxatio Ecclesiastica* of 1291; by 1324, however, much of its land had been drowned, and the survey of alien priories in that year explains that this is the reason why a reduced value is being put upon it.[35] A century later, Goldcliff parish church had been half-undermined by the sea, and a new site was chosen inland. The fourteenth- and fifteenth-century churches elsewhere on the Levels—St Bride's, Peterston and Marshfield to name three—were also built a safe distance inland.[36]

Drainage was undertaken to improve the land for pasture and to prepare it for tillage, and was carried out in conjuction with embankments along the coast and the natural watercourses which traverse the Levels. The Goldcliff monks dug a drain which is still known today as the Monks' Ditch—it was in some measure, perhaps, a replacement for the Roman work described above;[37] and an *inspeximus* of 1307 refers to a charter of Gilbert, Earl of Pembroke, *c.* 1140–50, in which the Tintern monks had received land on Magor moor divided *per fossas*, 'by ditches'.[38] This is the earliest reference to medieval draining on the Gwent Levels. Ninety years later, in 1245, Tintern was given permission by Walter, Earl of Pembroke and Lord of Striguil (or Chepstow), to enclose the land of its Moor Grange at Magor by a ditch, and to make consequential arrangements within. The same document also stipulates that the keepers of the mill at Aberweythel—on the coast—should assume responsibility for the watercourses running *per mediam albam waldam*, 'through the middle of the white *walda*', so that the monks should not suffer from any defect in it, or in the *gutta et clusa*, 'the gout and the sluice', which controlled the outlet to the Severn.[39] The reference is to an embanked watercourse known today as Whitewall Reen, probably from the light colour of the upcast, when fresh—a neighbouring reen, Blackwall, was perhaps cut down into the peat. The mention of a sluice is interesting at a mid thirteenth-century date, and as in the case of the 'Englishmen's Wall' on Margam's Afan Marsh was probably constructed by specially-engaged craftsmen. It is possible, too, the mill was a tide-mill; its position at what was, or was to become, one of the several little landing-places along the Gwent coast is indicative firstly of a considerable acreage of corn-growing land on the Levels, and secondly of a river-borne trade, perhaps with Bristol or some other centre. J. G. Wood pointed out many years ago that the name *Aberweythel* incorporates the Welsh words *aber*, 'mouth of,' and *gweithiau*, 'works',[40] it is apposite enough to describe the structures at the end of the reen.

In conclusion, let us glance at what was actually achieved at Tintern Abbey's Moor Grange within fifty years of the last-mentioned charter. In the *Taxatio Ecclesiastica* of 1291, we find that its land comprised 50 acres (20.24 ha) of meadow, valued at twice the normal rate for rough pasture in other holdings and therefore properly drained and managed, as well as anything in the abbey's possessions; and also two carucates or measures of ploughland, perhaps about 240 acres (97 ha) together. The valuation was, likewise, half as much again as the arable on other granges, and it was clearly rich and well looked-after.[41] In sum, we need have little hesitation in claiming that even in Roman times arable farming of well-situated and well-prepared alluvial land was practised, and contributed to the wealth of villas such as those at Wemberham and Kingsweston.

REFERENCES

[1] G. C. Boon, *Monographs & Collections* (Cambrian Archaeol. Assn.), **1** (1978), 1–24.
[2] Sir H. Godwin, *Journ. Ecology*, **31** (1943), 219.
[3] S. Locke, *Mon. Antiq.*, **3**, 1 (1970–1), 1–16.
[4] A. B. Hawkins in *Marine Archaeology* (ed. D. J. Blackman, 1973), 67–87.
[5] D. Lilly and G. Usher, *Proc. Univ. Bristol Spelaeological Soc.*, **13**, 1 (1972), 37–40.
[6] V. E. Nash-Williams, *Bull. Bd. of Celtic Stud.*, **14**, 3 (1951), 254–5 ['Redwick,' wrongly].
[7] G. C. Boon, *Mon. Antiq.*, **2**, 3 (1967), 121–7.
[8] S. G. Nash, *Proc. Som. Archaeol. & Nat. Hist. Soc.*, **117** (1973), 91–101.
[9] See note 30.

[10] H. C. Smyth-Pigott, *Proc. Som. Archaeol. & Nat. Hist. Soc.*, **31** (1886), pt. 1, 19, 23; H. M. Scarth, *ibid.* pt. 2, 1–9; R. C. Reade, *ibid.*, 63–73. The remains are 90 to 120 cm below grass (Prof. D. T. Donovan and A. M. ApSimon, fieldwork, 1962). The surviving finds are in the Woodspring Museum, Weston-super-Mare. The latest coin is of Constantine I, *RIC* Siscia no. 235, mentioned as 'Constantius' in the report cited and in V.C.H. *Som.* i, but cf. Scarth, *Proc. Soc Antiq. London.*, ser. 2, **11** (1885–7), 31–2.

[11] G. C. Boon, *Trans. Bristol & Glos. Archaeol. Soc.*, **69** (1950), 5–58.

[12] K. Branigan, *ibid.*, **92** (1974), 82–95.

[13] A. D. Berrington cited by Watkin, *Arch. Journ.*, **37** (1880), 137, 'found with bones' [unspecified]; Oct. Morgan, 'Goldcliff and the Roman inscribed stone . . . ,' *Caerleon & Mon. Antiq. Assn. Papers* (1882); cf. Scarth, *loc. cit.* note 10 above for the error. Cf. also J. K. Knight, *Mon. Antiq.*, **1,** 2 (1962), 34–6.

[14] *The Roman Inscriptions of Britain,* i (1965), no. 395.

[15] G. C. Boon, *Isca* (1972), 64–6.

[16] Quoted by Dorothy Sylvester, *The Rural Landscape of the Welsh Borderland* (1969), 395. For detail see also Margaret Davies, *Trans. Cardiff Nats. Soc.*, **85** (1958), 5–15.

[17] The largest deposit occurred on the site of Uskmouth B power-station, at Nash on the western tip of the Caldicot Level: Cefni Barnett, *Mon. Antiq.*, **1,** 1 (1961), 12–13.

[18] *Cal. I.P.M.* v (1908), 335 no. 358.

[19] The earliest (Flavian) surface was marked by a band of sedge and traces of burnt wooden buildings set between gravelled paths. This layer was separated from the third-century hard-standings by 20 cm or so of clean alluvium.

[20] 6 Hen. VI cap. 5. The effect of various facultative and temporary statutes was made perpetual by 3 & 4 Edw. VI cap. 8 (1549), which governed the Commissions until the Drainage Boards were set up in 1931. The date of the earliest Commission for Gwent is unknown. The earliest minute-book (in the County Record Office) is for 1692–5 and is labelled 'No. 7' (presumably of a series).

[21] Robert Ricart, *The Maire of Bristow is Kalendar* (ed. L. Toulmin Smith, Camden Soc., 1874), 46.

[22] Oct. Morgan (*loc. cit.* note 13) prints copious extracts from a chap-book, *Lamentable Newes out of Monmouthshire* (1607). There is a connected series of such publications. Cf. E. Green, *Proc. Som. Archaeol. & Nat. Hist. Soc.*, **24** (1879), 53–4; F. J. North, *Sunken Cities* (1957), 87–94.

[23] Welsh Agricultural Land Sub-Commission, *Mon. Moors Investigation Report* (1955), 32. Another tablet in Kingston Seymour church, near Clevedon, is apparently set about 60 cm higher, cf. M. Williams, *The Draining of the Somerset Levels* (1970), 88, note 2. The text, in C. E. P. Brooks and J. Glasspoole, *British Floods and Droughts* (1928), 97–8, reads: 'Jan. 20th 1606 and 4 Jas.I an inundation of the Sea Water by overflowing and breaking down the Sea banks happ[d] in the Parrish of Kingston-Seamore and many others adjoining by reason whereof many persons drowned and much Cattle and Goods were lost, the water in the Church was five feet high and the greater part lay on the Ground about 10 days.' Possibly the tablet has been moved; but an absolute coincidence of levels is hardly to be expected across the Bristol Channel. Since the measurements taken for the Sub-Commission at Goldcliff and Peterston are the same, great care has obviously been taken to replace the tablets there in the original positions whenever the buildings have been refurbished.

[24] A. W. Wade-Evans, *Vitae SS Britanniae et Genealogiae* (1944), 182–3, s11.

[25] Wm. Camden, *Britannia* (1607 ed.), 488. The preceding quotation, *Gent. Mag.*, **32** (1762, July), 306–8, probably from *Gods Warning to his People of England . . . by the late Overflowing of the Waters . . .* (1607)

[26] I am much obliged to Mr G. McLeod, of the Usk River Division of the Welsh National Water Development Authority, for kindly giving me access to the records of tidal measurements.

[27] M. Williams, *op. cit.*, 18–24, cf. 41–44.

[28] *Vitae SS Britanniae,* 210–13, s13.

[29] G. T. Clark, *Cartae et alia Munimenta quae ad Dominium de Glamorgancia pertinent* (6 vols., 1910), no 50.

[30] L. Dudley Stamp, *Geog. Journ.*, **93** (1939), 498, 501.

[31] The Cistercians were exempt from payment of tithes, an added inducement: cf. G. Constable. *Monastic Tithes* (1964), 241–2. The purpose was probably to assist new and struggling foundations.

[32] *Cartae,* nos. 565 (*c.* 1196–1270); 1595 (early 13th century); 710, 809, 810 and 921 (*c.* 1271–1331); 1571 (13th century). T. Gray, *Journ. Brit. Archaeol. Assn.,* n.s. **9** (1903), 166, alludes to a wall of masonry ('*Gwal Saeson*') as then still existing in considerable part, parallel with the Afan. The stonework would doubtless be later than the period which we are considering.

[33] *Cartae,* no. 710 (*c.* 1271–1331).

[34] *Cartae,* no. 445 (*c.* 1226–9).

[35] For Goldcliff Priory see D. H. Williams, *Mon. Antiq.,* **3,** 1 (1970–1), 37–54, esp. 44, 51–2.

[36] The villages would of course be earlier. Cf. J. K. Knight, *Mon. Antiq.* **3,** 1 (1970–1), 17–19, on Anglo-Norman settlements and *-ton* names.

[37] It runs north-east of the knoll, whereas the stone was found to the west.

[38] *Cal. Charter Rolls* iii (1908), 88–97.

[39] *Ibid.,* 104–5.

[40] J. G. Wood, *The Manor and Mansion of Moyne's Court* (1914), 103–5.

[41] *Taxatio Ecclesiastica Papae Nicholai* (1802), 282b; D. H. Williams, *Mon. Antiq.,* **2,** 1 (1965), 16.

# The Evolution of Romney Marsh: a Preliminary Statement

Professor B. W. Cunliffe, F.S.A.

The area commonly known as Romney Marsh (fig. 17) consists of 100 sq. miles (259 sq. km) of marshland and shingle projecting out into the English Channel between the towns of Hastings and Folkestone. Most of it lies in the county of Kent but the south-western corner comes within the boundary of East Sussex. Strictly the region referred to as Romney Marsh comprises the marshes of Romney and Walland (divided from each other by a medieval drainage earthwork, the Rhee Wall), together with Denge Marsh, the Levels of Pett, East Guildeford, Broomhill and New Romney, and a series of shingle beaches of which the greatest and most dramatic is Denge Beach terminating in the headland of Dungeness, now crowned by a nuclear power station. The Marsh is a world apart—a constantly changing landscape moulded by its rivers, by the sea and by man.

At its simplest the Marsh can be said to comprise an expanse of variegated sediments 40–100 ft. (12.2–30.5 m) thick occupying a shallow bay cut into Wealden Clay and Hastings Beds and enclosed on the landward side by degraded sea cliffs which run from Hythe to Winchelsea in an arc broken only by the valleys of the rivers flowing out from the Weald. The processes by which the sediments accumulated are complex and interwoven. Not surprisingly the evolution of the Marsh has been the subject of lively and sometimes vitriolic debate for three centuries.

There would be little value here in discussing in detail the theories and beliefs of the early writers (e.g. Holloway 1849). Suffice it to say that the first serious attempt to understand the early history of the Marsh was made by the local engineer James Elliott who collaborated with Charles Roach Smith in his excavations at Lympne in 1850 and contributed a chapter to Smith's report (Elliott 1852, 37–45) in which he outlined his views on the changes to which the Marsh had been subjected in the Roman period. Thereafter for almost 60 years the story of the

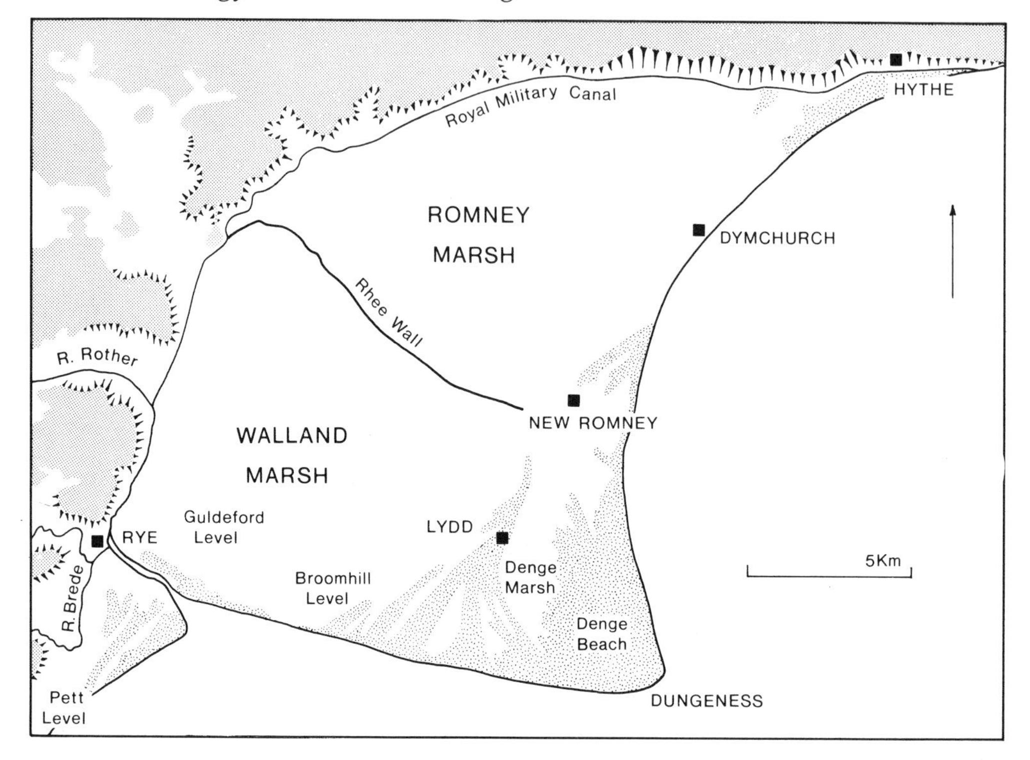


Fig. 17. Romney Marsh.

Marsh formed a constantly recurring theme in the interminable discussions which surrounded the invasion of Julius Caesar. The outpourings of this academic interlude were admirably, if somewhat acidly, summarised with full supporting bibliography by Rice Holmes in his *Ancient Britain and the Invasion of Julius Caesar* (1907, 532–52).

The 1930s saw major advances in knowledge. In a series of papers, using the topographical evidence contained in several Saxon Charters, Gordon Ward attempted to build a picture of the Marsh in the pre-Norman period (Ward 1931a; 1931b; 1933a; 1933b; 1936). Although some of his conclusions are now in dispute his contribution to the debate was invaluable.

It was during this period that W. V. Lewis published his seminal paper on the formation of Dungeness (Lewis 1932). Meanwhile the broad outlines of the history of the Marsh were considered afresh by Gilbert (1933) and Homan (1938); Teichman Derville (1936) contributed an historical commentary, while Wallenberg's study of place names (1931) added another dimension to the debate. The decade of research culminated in a paper by Lewis and Balchin, read to the Geological Society in March 1940, in which they presented the results of a programme of surveying involving the careful levelling of the beach ridges of Dungeness. From this work they were able to relate the growth of the foreland to changes in the height of mean sea level. Their paper, and the published discussion which ensued (Lewis and Balchin 1940) cleared the way for the complete reassessment of the evolution of the Marsh. Of particular significance was the demonstration by Ward (in the discussions *ibid*. 281–2) that the Rhee Wall, formerly held to be Roman, was in fact medieval.

The work on Dungeness provided sound geomorphological data to which was added the evidence obtained from bore holes and from observations made by the staff of the Geological Survey (summarised in Smart 1964; 1966a; 1966b), but no adequate overall consideration was available until work began in 1954 on a systematic soil survey of the entire area of Romney Marsh at the instigation of the Soil Survey of England and Wales (summaries in Green and Askew 1958a; 1958b; 1959; 1960). The final report, published fourteen years later (Green 1968), together with a detailed soil map at a scale of 1:25,000, forms the essential basis for all future work. Without it it would have been impossible to write this paper.

## *The Processes Involved in the Development of the Marsh*

The dynamics of marsh formation in the Romney area are complex. Four distinct but closely interrelated processes can, however, be isolated.

### (a) *The growth of the beach*

The shingle beach which developed (and still continues to develop) around the seaward fringe of the Marsh was a controlling factor of considerable significance. The shingle spit formed by a process of long-shore drift. Beach pebbles, created principally from the erosion of the chalk cliffs of Sussex, were moved eastwards along the coast by the predominantly west–east currents (Lewis 1932). From Fairlight Head the beach built out across the edge of the shallow lagoon (within which the Marsh was to form) until it met the coast again in the neighbourhood of Hythe. The considerable volumes of fresh water which poured out of the Weald had to penetrate this bar in order to reach the sea. The positions of their outlets, which changed through time, affected the growth of the bar. The dynamic relationship between the scouring of the river mouths by fresh water and the constant movement of shingle by the tides ensured that, while the shingle beaches grew by accretion in some areas, in others older beach lines were actively eroded.

### (b) *The rivers*

Three rivers, the Brede, Tillingham and Rother, flow through the Marsh into the sea. Much of the silt load which they carried was deposited in the area behind the shingle beaches. When velocity was low (in times of high sea level) the deposition of silt would have been considerable: in periods of lower sea level the regraded rivers would have carried the greater proportion of their load to the sea. The rivers were a vital factor in the formation of the shingle bar for the reasons noted above.

### (c) *Changes in mean sea level*

Variation in mean sea level will have affected the Marsh in a number of ways. A rise in sea level would have had two principal effects, causing the heights of the storm beaches to increase (Lewis and Balchin 1940) and forcing the rivers to regrade, this second factor leading to a period of extensive sedimentation in the protected area behind the beaches. If sea level continued to rise then marine transgression would have ensued leading to sedimentation of a different type. Conversely a fall in sea level would have led to the creation of lower beach ridges while the rivers would have cut deep channels through the silts allowing areas previously flooded to dry out.

The complexity of isostatic change has been fully considered in general terms elsewhere in this volume (pp. 1–23) while the pitfalls of dealing with the more recent period in south-eastern England have been carefully presented by Akeroyd (1972). More recently Tooley (1974) has demonstrated the complex nature of the minor fluctuations in sea level affecting the north-west coast of England in the last 9,000 years. If changes of this magnitude occurred in the south-east they will have been of considerable significance in the formation of Romney Marsh.

### (d) *The effects of man*

The interference by man with the natural drainage of the Marsh was a decisive factor. The colonisation of large tracts of Romney Marsh proper in the late Saxon period, with the attendant small scale drainage work which must have accompanied it, cannot have failed to have affected the landscape. Drainage brought with it a lowering of ground level as the result of the drying out and compaction of the underlying sediments (Green 1968, 24–27)—a factor which will have caused further drainage works and embanking to be undertaken. In the south, in the area of Walland Marsh, there is clear evidence, both from documentary sources and in the form of a complex system of dykes, of the process of reclamation which took place throughout the medieval period. Finally man has arrested the natural changes taking place along the shingle beach by artificially strengthening certain sections of it. The most dramatic example of this is the creation of the Dymchurch Wall described in fascinating detail by the engineer responsible (Elliott 1847).

## *The Principal Stages in the Formation of the Marsh*

It is not proposed to discuss here in any great detail the changes which have taken place in the formation of Romney Marsh. To do this would first require an extensive programme of core-boring, combined with the necessary scientific analysis and radiocarbon dating, and a reassessment of all the documentary evidence together with a systematic topographical field survey. This work has hardly begun. There is, however, sufficient evidence contained in the scattered

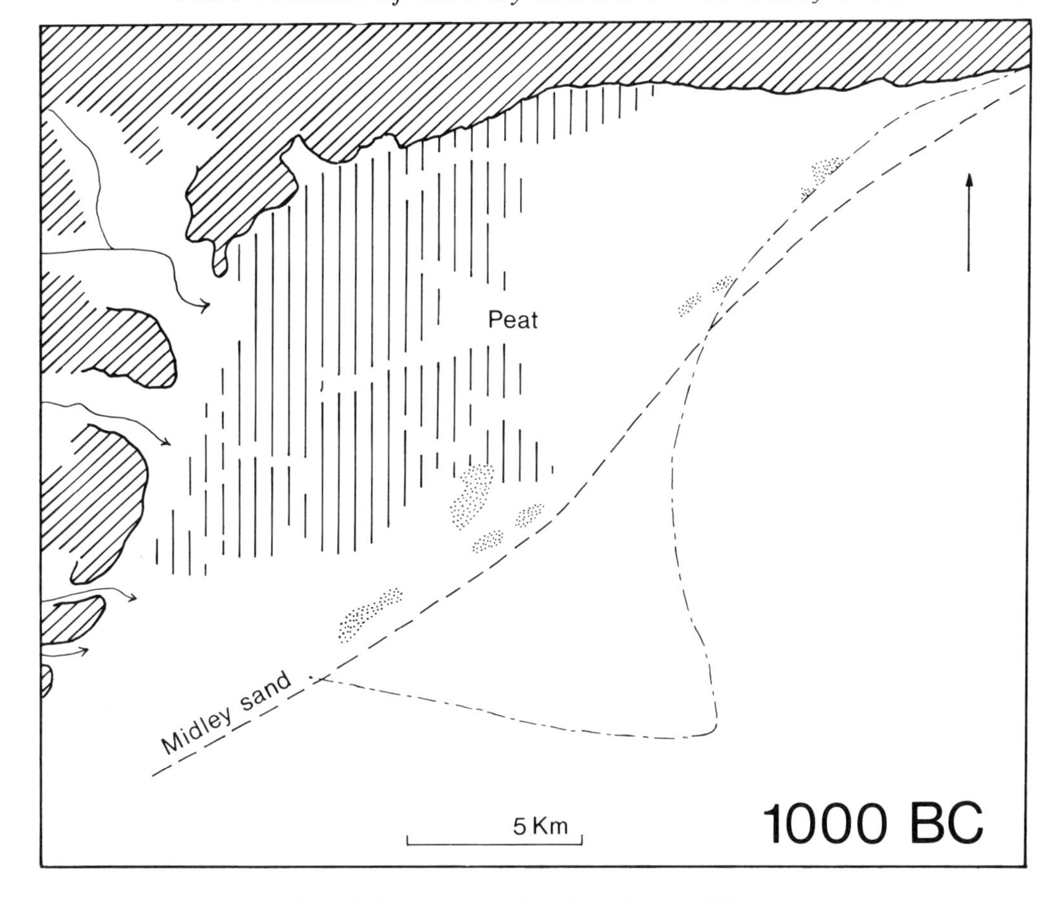


Fig. 18. The Romney Marsh region *c.* 1000 B.C.

literature to allow the main stages in the evolution of the Marsh to be outlined. Such an attempt is offered here as a corrective to the misconceptions which still pervade the general archaeological literature and to focus attention on the very considerable potential of the region for landscape studies.

*The pre-Roman situation* (fig. 18)

The gradual rise in sea level, which ensued with the onset of the Neothermal period, after 10,000 B.C, led to the formation of a wide shallow bay as the encroaching sea cut deeply into the soft Wealden rocks between Fairlight Head and Folkestone, creating the cliff, now considerably degraded, which today marks the landward limit of the Marsh. As the sea level rose the cliff line advanced into the Weald while at the same time the detritus derived from the erosion was deposited over the floor of the bay to the edge of the old submerged sea cliff, the line of which is now represented by the 10 fathom contour.

At some stage, partly as the result of isostatic readjustment and partly because the bay was becoming choked with debris, accumulation outstripped erosion. The bay now became a shallow lagoon choked with sandy shoals and fringed along its seaward edge by a sand bar represented by the Midley Sand. This Midley Sand, with high coarse sand component and little silt (Green 1968, 13–14), must represent a series of sand spits or dunes which once emerged from the sea. Wherever the sand has been examined in relation to other deposits it can be shown to lie on the older detritus and to precede both the shingle beach and the alluvial and organic deposits of which the upper levels of the Marsh are composed.

The deposition of the Midley Sand seems to have been followed by the accumulation, on its seaward fringe, of a series of shingle beaches which now lie to the west of Lydd and follow the SW–NE alignment of the sand bar. These beaches, or shingle spits, created by long-shore drift from pebbles derived principally from the Sussex coasts, represent the beginning of a continuous process of beach formation which even today is changing the configuration of Dungeness.

The next depositional phase within the bay is marked by the accumulation of a thickness of Blue Clay containing shells of *Scrobicularia*—an organism which can flourish only below mid-tide level. Its presence therefore suggests the prevalence of marine conditions within the lagoon resulting from a period of subsidence and flooding. Green (1968, 14) has suggested that the Blue Clay may be stratigraphically related to the Buttery Clay of the Fenlands which represents a period of marine transgression dated broadly to the late third and early second millennia B.C.

The Blue Clay is followed by an extensive layer of peat containing, in its lower levels, the trunks and roots of oak, birch and other trees. The fact that in several areas the trees can be shown to be rooted in an ancient ground surface is a clear indication of a period of low sea level when forest colonised the silted up lagoon. The peat is at its thickest in the Dowels, east of Appledore, where exposures 6 ft. (1.82 m) thick have been noted above Blue Clay, the top of which is at *c.* 5 ft. (1.52 m) below O.D. To the south-east the peat thins out as it approaches the elevated ridges of Midley Sand and the earliest shingle beaches. The forest landscape will have varied from place to place. On the evidence of macroscopic remains oak and birch predominated but hazel and alder have been recorded (Drew 1864, 15). The effects of man upon the forest are still largely undefined but the claim by Dowker (1897, 212) that some of the stumps bore axe marks, and the discovery, in the submerged forest off Pett Level, of a worked flint flake (Milner and Bull 1925, 320), are reminders that man was now beginning to play a formative part in the development of the Marsh.

The dating of the 'Forest period' rests upon two radiocarbon assessments:

| | | |
|---|---|---|
| Peat from Higham Farm (977308) | NPL23 | 1070 ± 94 bc. |
| Tree trunk from Court Lodge (032243) | NPL24 | 1390 ± 92 bc. |

Although too much reliance cannot be placed on isolated dates of this kind they indicate that forest growth was probably taking place in the second half of the second millennium.

Sometime during the first millennium B.C. a marine transgression occurred giving rise to a layer of fine alluvium which everywhere seals the peat of the 'Forest period'. No doubt the processes leading to the inundation were complex. In the early stages pools and bogs would have formed in the forest as the fresh water from the Wealden rivers, finding access to the sea more difficult, would have ponded up in low lying areas behind the shingle beach line. As conditions worsened sediments brought down by the rivers would have accumulated above the peat while in the estuaries, with the increasing inroads of the sea, brackish water conditions would have begun to extend further inland. Under such conditions slight alterations to river courses or to tide level would have affected large areas of countryside. Although the detail of these changes is far from clear the situation at the beginning of the first millennium A.D. would appear to be that the western part of the lagoon was occupied by a slightly elevated marshland, probably susceptible to periodic flooding, while the eastern part was an open tidal estuary with its mouth in the vicinity of Hythe.

*The Roman and early Saxon periods* (figs. 19 and 20)

An indication of the situation in the Roman period is given in fig. 19. At first sight the map might appear to reflect an over-optimistic precision unsupportable by evidence but the soil survey work published by Green (1968) contains most of the data upon which the figure is based. Crucial to the problem is the chronological difference implied in the distinction made between two types of marshland, the older which has become decalcified, and the newer which still retains a significant calcium content (*ibid*. 23–29). It is the old (decalcified) marshland which here concerns us. More detailed analysis has shown that this marshland was dissected by complex systems of creeks (represented now by slightly elevated ridges) all of which appear to have flowed in a north-easterly direction towards the area infilled by the later (or calcified) marshland. Figure 19 was created by plotting the decalcified marshland (shown in light stipple) on the basis that it was once dry land traversed by a multitude of creeks leading into a wide estuary, the principal outlet of which lay in the vicinity of Hythe. Green (1968, fig. 14) allows the possibility of another outlet at Romney. While this may have been so there is no positive evidence for or against such a view and the simpler hypothesis is preferred here—that the shingle beach was continuous, except for the Hythe outlet, and that all the creeks were part of a single drainage system. If so, we can assume that all three Wealden rivers, the Brede, Tillingham and Rother, flowed into the single estuary.

The volume of fresh water which must have passed through the constricted mouth of the estuary was considerable. All the time that the flow was maintained the drift of shingle eastwards was prevented from blocking the exit and instead a cuspate headland was created, its fulls (shingle ridges) curving back at right angles to the main beach. They were noted by Elliott during his construction of the Dymchurch Wall (Elliott 1847, 42–3) and the tips of the fulls still survive though heavily mutilated by modern gravel extraction.

The dating of the topographical situation mapped in fig. 19 depends upon the location of Roman material and its relationship to marsh stratigraphy. At Jesson Farm, St Mary-in-the-Marsh pottery of the late first century B.C. or early first century A.D. was discovered on an old land surface (Green and Askew 1960).

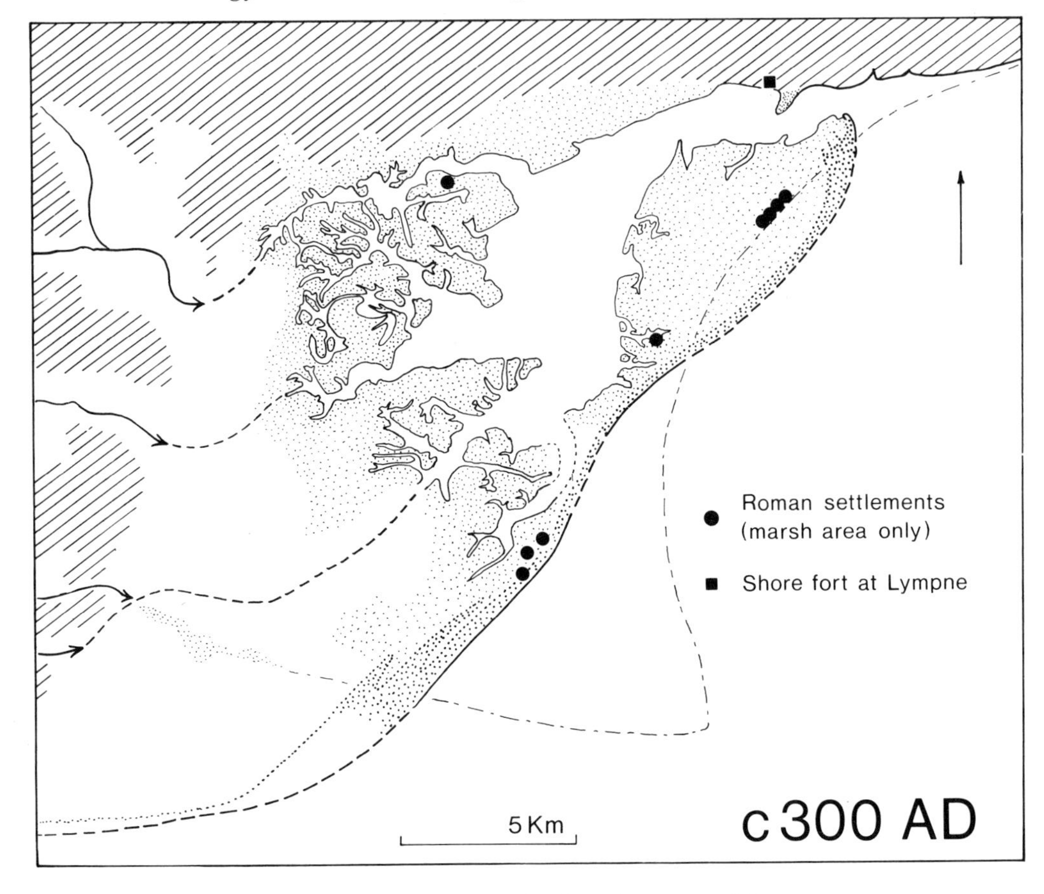


Fig. 19. The Romney Marsh region *c.* A.D. 300.

Nearby, close to St Mary's Bay, a land surface scattered with Roman occupation debris was found to be dissected by a minor creek (now a creek ridge), the loamy filling of which contained a lump of the eroded Roman ground surface (Green 1968, 113–4). The implication would seem to be that the creek had cut through a habitation area of the early Roman period. Shells collected from the filling of a creek in the same system south west of Old Romney were dated by radiocarbon analysis to A.D. 400 ± 120 (NPL25) (Smart 1964) suggesting that, at this point at least, silting was taking place in the late Roman or early Saxon period.

One further point is of some relevance here. During the construction of the Dymchurch Wall extensive areas of Roman occupation material were exposed. The published descriptions (Dunkin 1844 and Isaacson 1846) indicate that in addition to normal domestic rubbish quantities of what appears to have been briquetage were recovered together with a number of cremation burials. Clearly the area of old marshland protected by the shingle beach (now the St Mary's Bay to Dymchurch area) was the scene of considerable activity in the early Roman period: salt extraction possibly accompanied by pottery manufacture was being carried out, while the cremation cemetery is indicative of a settled community.

Roman occupation, though apparently on a less extensive scale, has also been recorded on the old beaches behind Lydd (Jones 1953).

Additional evidence of the situation in the Roman period has been recovered from the recent excavation of the Roman fort at Lympne (Cunliffe, forthcoming). Here it was possible to demonstrate the existence of a shore line, probably of Roman date, 6 ft. (1.8 m) below the present level of the marsh. The siting of the fort, and its *Classis Britannica* predecessor, on the shore commanding the narrow entrance to the estuary, makes good strategic sense. From such a position it would have been possible to control all the shipping entering the harbour and to oversee the riverine and coastal transport of iron mined and extracted in the Weald. The actual location chosen, immediately to the west of a substantial sand dune, overlooks the best sheltered anchorage in the vicinity.

If it is possible to outline the general state of the Marsh in the Roman period with some degree of accuracy it is less easy to define the changes which were taking place during this time. The first-millennium B.C. rise in sea level which submerged the forested landscape must have been reversed to some extent to allow large areas of alluvium in the lee of the shingle beach to dry out sufficiently to be colonised in the first and second centuries A.D. Thereafter the sea level rose again and it was probably as a result of this that the creek system shown on fig. 19 reached its greatest extent as the higher tides cut more deeply into the old marshland. All this time long-shore drift shaped the protecting shingle beach, the height of the individual fulls rising with the sea level (Lewis and Balchin 1940).

At some stage in the late Roman or early Saxon period it would seem that the shingle barrier was breached at two additional places: in the vicinity of Romney and some miles to the south of Rye. Although the breaching may in part have been due to the rise in sea level, a more direct cause should be sought in the dynamics of shingle spit formation. From an early stage local tidal conditions have encouraged the deposition of shingle in such a way as to create a headland (now Dungeness) which has a constant tendency to grow in an eastwards direction. Much of the material for the head is derived from the erosion of the more westerly parts of the earlier beaches (Lewis 1932). In such a process the stage must have been reached when the last of the shingle was removed at the western extremity of the beach and the sea broke through, flooding the low lying marshland behind. The extent of the flooding will have been affected by the degree to which the level of the decalcified marshland had subsided through the compaction of the alluvial sediments and the shrinkage of the peat beneath. Since it would seem that the peat was thickest in the west it is reasonable to assume a greater degree of compaction here, and hence a lower overall level, giving rise to widespread inundation. A further result of such a process would have been that the rivers Brede, Tillingham, and possibly the southerly branch of the Rother were captured and now flowed southwards to the sea, the not inconsiderable volume of fresh water scouring their new outlet and keeping it free from further silting. The precise extent of the flooding cannot yet be plotted but the relic creeks planned by Green (1968, fig. 14) which can be shown to post-date the creek ridge system and pre-date the earliest innings are presumably of this period (fig. 20).

The effect of the capture of the southern rivers on the original (now northern) estuary must have been dramatic since the volume of fresh water, which had previously kept the mouth clear, was considerably reduced. This factor combined

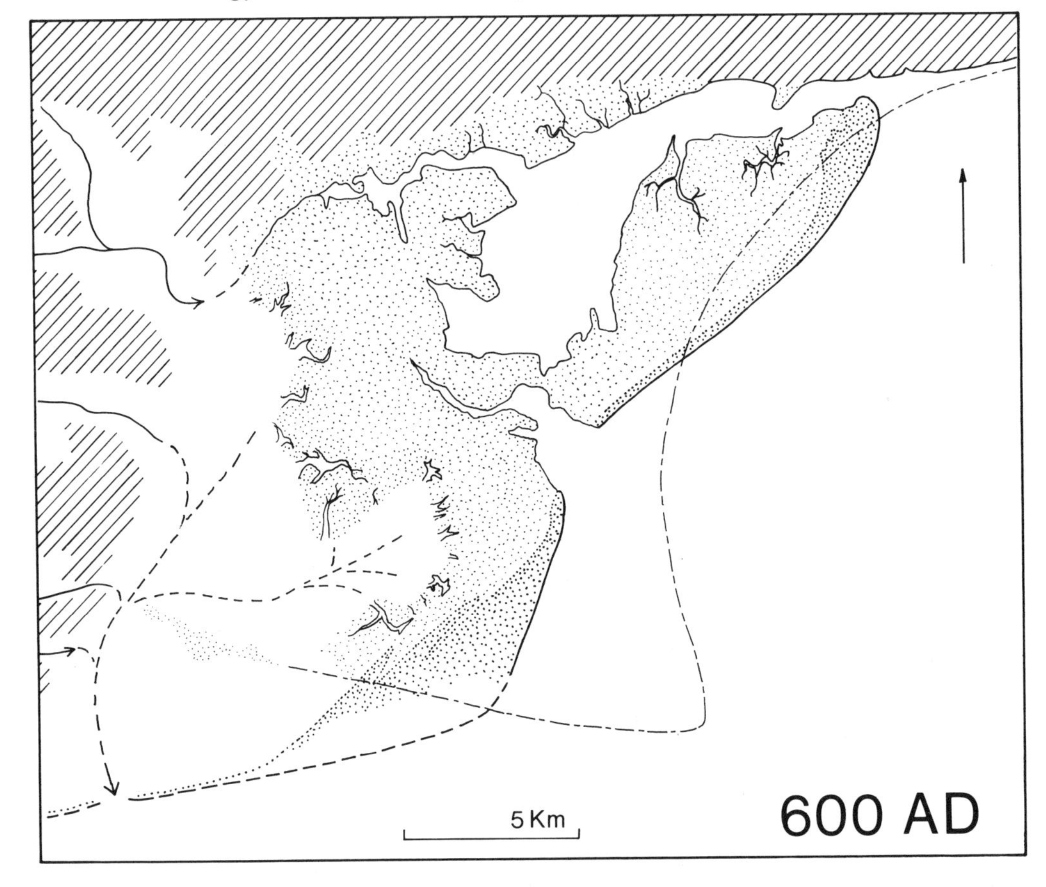

Fig. 20. The Romney Marsh region *c.* A.D. 600.

with the higher sea level will have caused increased sedimentation within the estuary, not least because there was little to prevent the end of the gravel bar from building out further across the narrow outlet. The silting of the principal creek system noted above (with the radiocarbon date of A.D. 400 ± 120) was part of this process. It is significant that the one substantial creek which seems to have been kept free from silt was that through which the main course of the Rother flowed.

The second break to have occurred in the gravel bar at this time was in the vicinity of Romney. The reasons for this are easy to appreciate. With the main (proto-Dungeness) headland building out, the force of the currents rounding the head were directed back against the beach in the Romney region with the result that much old shingle was eroded and the sea soon broke through creating a small estuary. Unlike the northern and southern estuaries, at first it had no river to feed it but it would no doubt have formed the principal element in a localised drainage system, taking surface run-off from the central area of the Marsh. This factor, together with erosion caused by tidal surges seems to have extended the creek further and further inland until at some stage it must have reached the Rother, tapping part of the water which had previously flowed through the northern estuary. With the volume of water further diminished the northern estuary soon became clogged with silt (the calcified marshland on Green 1968, fig. 16), all that survived of the drainage pattern being a single creek system.

The complex processes outlined above spanned the period from the elevation of the marshland in the late first millennium, which was followed by colonisation in the early Roman period, to the final silting of the northern estuary some time in the seventh or eighth century by which time, as the charter evidence shows, the area was already dry land (below, this page). The sequences of events suggested here is based on an interpretation of the soil survey evidence presented by Green (1968). At best it should be regarded as an hypothesis which requires testing and refinement but the principal stages at last seem to be reasonably well established.

*The late Saxon period* (fig. 21)

By the eighth century the picture becomes much clearer, largely because a surprising number of land charters relevant to the area survive. These have been studied in some detail by Ward (1931a; 1931b; 1933a; 1933b; 1936) and while it must be admitted that a complete reconsideration of the evidence is long overdue, the broad outlines of his conclusions are likely to remain unaltered. In summary, several charters of the period A.D. 720–850 concerned with land apportionments between Ruckinge and West Hythe (the area of calcified marshland filling the northern estuary) mention, as a significant landscape feature, the river Limen. The river no longer survives but Ward thought that part of its course could be identified with Sedbrook Sewer, just south of Ruckinge. This view is accepted by Green and Askew (1959, 25) who have convincingly traced the rest of the course, as a meandering creek ridge followed in part by later roads, as far as West Hythe (fig. 21). That the creek served as part of the boundary of seven parishes is an indication of its one time significance. It can reasonably be assumed to have been the main artery of the drainage pattern of Romney Marsh proper at the time when the newly won marshland on either side was being divided and apportioned. Much of the area to the north of Limen creek was added to the parishes situated along the cliff edge while the greater expanse to the south became the parishes of Ivychurch, Newchurch and Eastbridge, the boundaries of which correspond closely to the edge of calcified (new) marshland soils: all three are mentioned in Domesday Book. Thus, taken together with the charter evidence, it can be fairly assumed that the reclamation and settlement of Romney Marsh proper must have been completed in the period from the mid eighth to mid eleventh century.

Less certainty attaches to Walland Marsh. The churches at Fairfield, Brookland and Midley show, as might be expected, that the southern fringes of the old (decalcified) marshland was settled in this period, but the land further south was almost certainly tidal salt marsh apart from the fringe behind Lydd Beach. Ward's attempt (1931a) to argue that the area now known as Denge Marsh, between Lydd and Denge Beach, was granted in a charter of A.D. 774 is based on dubious topographical arguments and carries little conviction in spite of Lewis and Balchin's (1940) acceptance of the theory—an acceptance which incidentally caused difficulties to their own hypothesis.

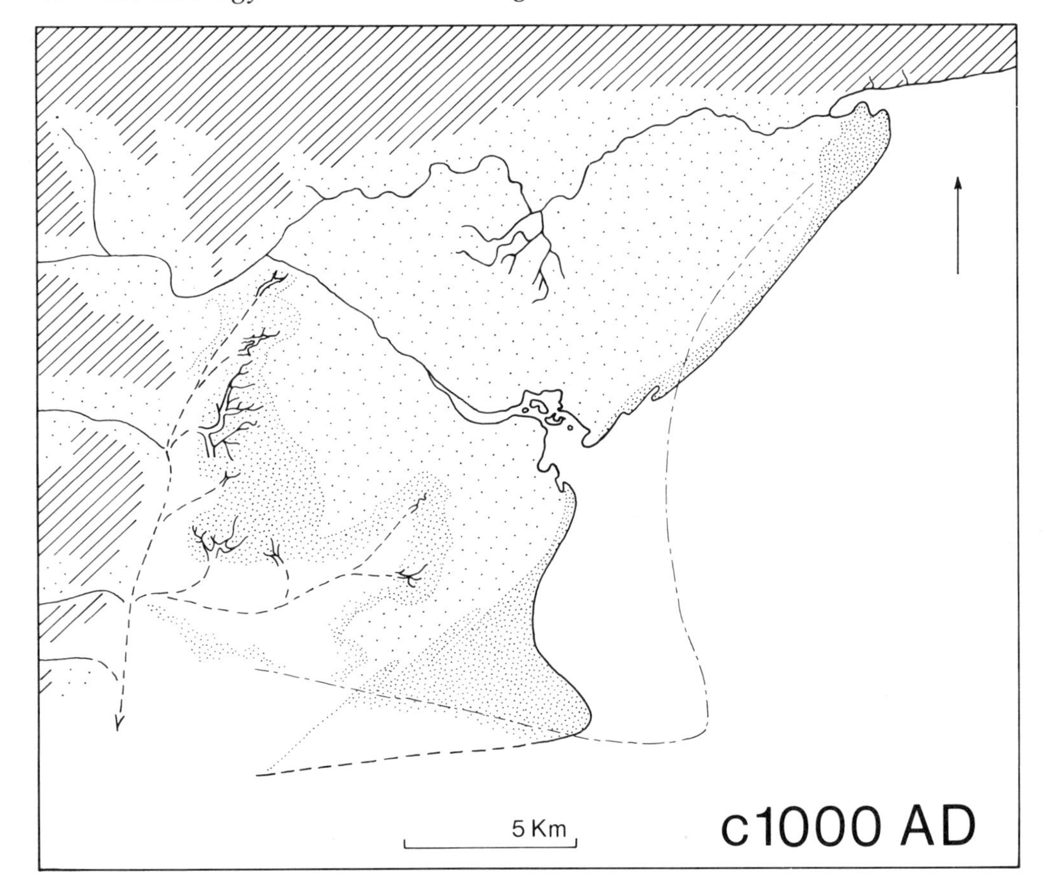


Fig. 21. The Romney Marsh region *c.* A.D. 1000.

If the sequence of events outlined above is accepted, then by the ninth or tenth century the river Rother probably had two outlets, a northern channel called the Limen reaching the sea near Hythe and a southern channel, also referred to in a charter as the Limen, emerging at Romney. The two other Wealden rivers, the Brede and Tillingham flowed into a wide estuary emptying to the south.

Against this geomorphological background we must briefly consider the two well-known contemporary descriptions of the area: the Anglo-Saxon Chronicle for A.D. 893 records that the Danes with 250 vessels 'sailed into the mouth of the River Limen in Kent, which floweth from the great wood that is called Andred; thence they towed up their boats four miles into that wood from the mouth of that river'. The description could, as Green has acknowledged (1968, 36) refer to the northern creek so long as it is assumed that it was navigable as far as Newchurch, but it could equally well apply to the Romney branch of the Limen which gives the appearance of having been wider and more accessible at this time. On balance the Romney branch would seem to be a better fit. The second description by Nennius, a tenth-century source, describes Lomnon (Limen) Marsh: 'in it are 340 islands with men living on them. It is girt by 340 rocks and in every rock is an eagle's nest and 340 rivers flow into it: and there goes out of it into the sea but one river which is called the Lemin (Limen).' So fanciful a description hardly bears critical examination but the overall impression given is of an extensive estuary with a narrow outlet to the sea—hardly a fitting description for the northern or Romney creek but one which might have reflected the conditions prevalent in the southern estuary. An acceptable alternative explanation would be that Nennius's description did indeed refer to the northern estuary but at an earlier stage in its development. The possibility, that all three outlets may have been known as the Limen, renders further detailed analysis unnecessary.

*The medieval period* (figs. 22–24)

From the beginning of the medieval period the history of the development of the Marsh takes on a greater precision. So extensive are the available data that only a brief outline can be given here.

Romney Marsh proper was by now extensively settled but the northern estuary may still have remained navigable, at the beginning of this period, as far west as the ancient sand ridge which lies to the west of West Hythe. The ridge is almost certainly the site of Sandtun mentioned in 732 (Ward 1931a) and pottery found in a sand pit confirms that a settlement existed here in the tenth or early eleventh century. The silting of the estuary and the growth eastwards of the shingle beach may well have caused the harbour settlement to migrate eastwards, first to West Hythe, then to Hythe and finally, after Hythe Haven had become clogged with shingle and silt at the end of the sixteenth century, to Folkestone. Without a flow of fresh water from the Limen to keep the harbour clear of sediment its decline was inevitable.

By the early medieval period much, if not all, of the water of the Rother was reaching the sea through the Romney creek (fig. 23), the sinuous course of which is readily distinguishable on the soil map, its filling characterized as 'Land type with mounds' (Green 1968, 39–40). The shingle beaches of Romney and Lydd protected the seaward flanks of a sheltered inlet on the south shore of which lay the town of Old Romney: New Romney occupied the northern beach. All the time that the Rother continued to flow through the channel the estuary was kept clear of silt and shingle but the subsidence of the land around Appledore (resulting from the compaction of the underlying peat) threatened to divert the Rother southwards into the southern estuary. To counter this a new canal, the Rhee Wall, was dug, to lead the waters of the Rother on a more direct course from Appledore to the estuary at Old Romney.

The date of the Rhee Wall has been widely debated. Green in a detailed discussion of the documentary and the soil evidence (1968, 37–42) argues that it was constructed prior to 1257 in which year a Patent Roll refers to the making of 'a certain new course' to channel water, recently diverted into the southern estuary, back to Romney. His contention is that the 'new course' extended the existing Rhee Wall eastwards past Old and New Romney to the sea, making a way for the water through the rapidly silting estuary. Thus three stages in the evolution of the system are envisaged:

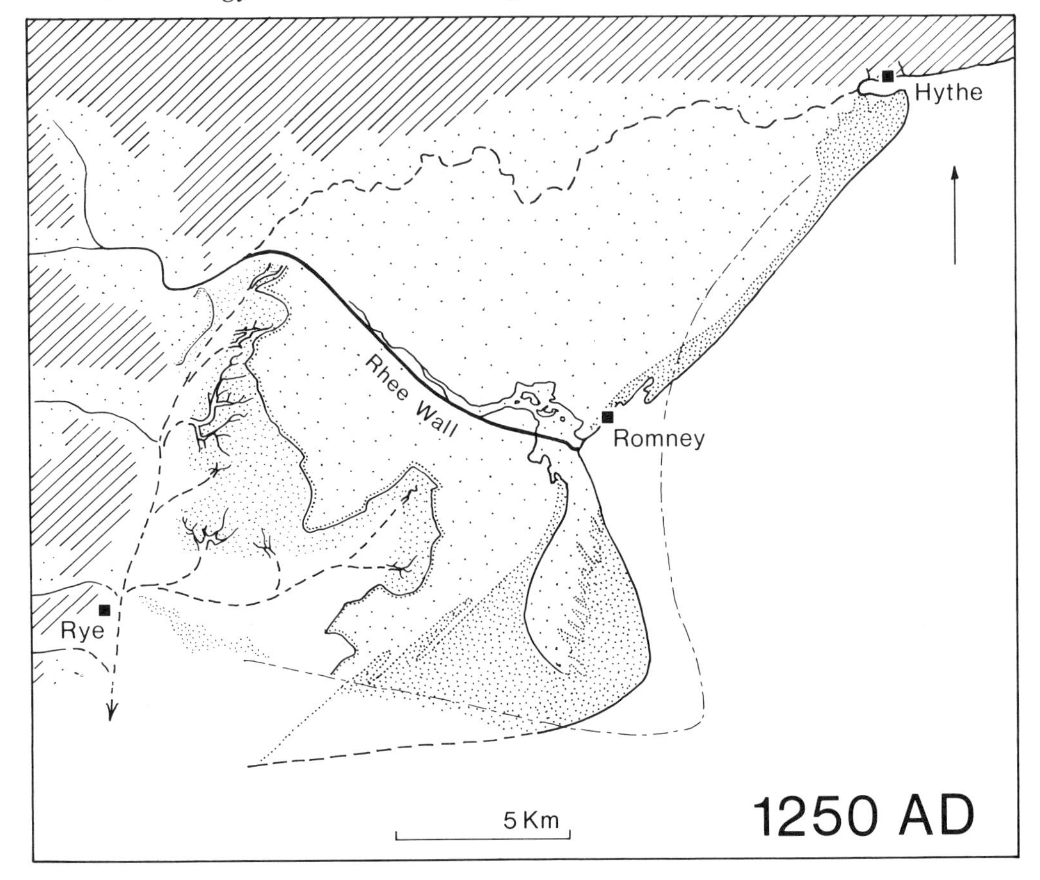


Fig. 22. The Romney Marsh region *c.* A.D. 1250.

(a) the original channel which had developed in the Saxon period
(b) the first canalisation from Appledore to Old Romney dating to the late twelfth or early thirteenth century
(c) the extension eastwards which was constructed in 1257.

The great storms of the thirteenth century had a dramatic effect on the formation of the marsh. In October 1250 exceptionally high tides destroyed ships and flooded more than 300 houses at Old Winchelsea, but the greatest of the storms was that of 1287. Old Winchelsea and Promehill (now Broomhill) were totally destroyed as the sea surged into the southern estuary, flooding all the marshland south of the Rhee Wall and finally diverting the Rother into the southern estuary, leaving Romney without the flow of fresh water to keep the harbour clear of sediment. Even the town of New Romney, perched high on its gravel beach, was flooded to such an extent that four feet of silt accumulated in the church of St Nicholas, the floor of which is at over 16 ft. (4.88 m) O.D.

Events of this kind must have left their mark on the shingle beaches of Dungeness. Lewis and Balchin (1940, 273) suggest that the highest of the shingle fulls

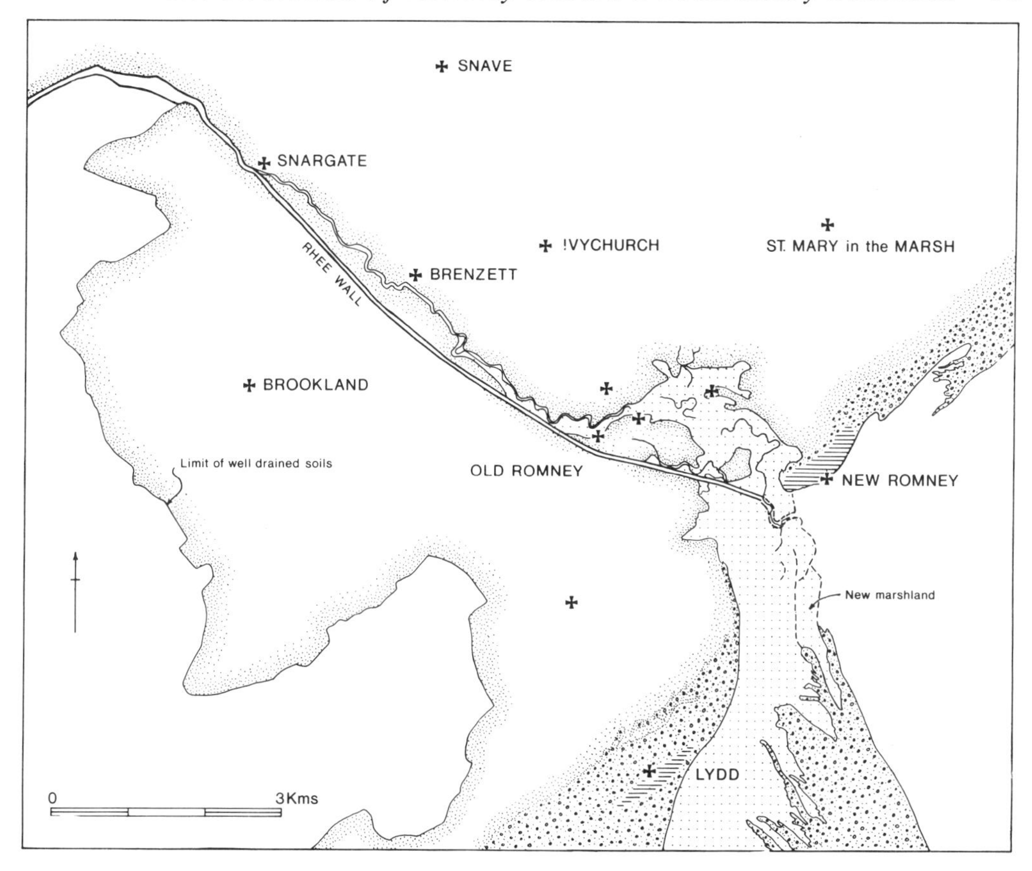


Fig. 23. Land type and settlement in the vicinity of Romney.

(20.65 ft. O.D. (6.2 m)) should be equated with the events of 1287. This line, marked as the shore line on fig. 22, shows the extent to which the headland must have advanced, its rapid growth being caused by the exceptional storms of the preceding decades.

The advance of the headland and the loss of the Rother marked the end of Romney as a port. After the storm rapid silting took place, extending from behind the newly formed shingle beaches, in the area of Denge Marsh, out across the original estuary (fig. 23). The situation is vividly summed up by Camden writing in 1586. After describing the great storm of 1287 and the transference of the Rother to its new outlet near Rye he concludes 'then it (the sea) began by little and little to forsake this town, which has decayed by degrees ever since and has lost much of its populousness and dignity'.

The area of the southern estuary, now Walland and Guildeford Marshes, is characterised by Green as 'Land type with common creek relics' (1968, 30–34). It is new (calcarious) marshland and bears extensive evidence of the process of reclamation to which it was subjected in the medieval period. The overall plan of the innings walls are shown on fig. 24 from which the complex piecemeal nature of

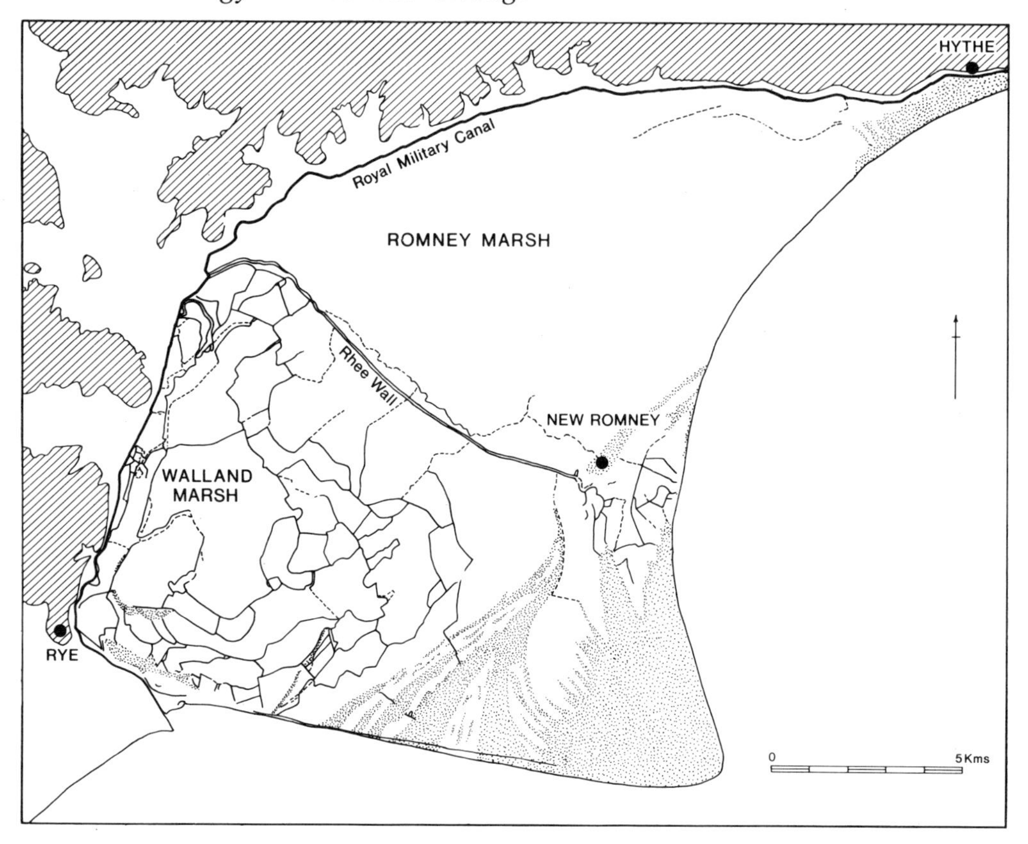


Fig. 24. Innings boundaries on Romney Marsh.

the reclamation will be evident. Nor should we expect the process to have proceeded uniformly: the great storms of the thirteenth century are known to have flooded vast tracts of previously drained land which would have had to be won back from the sea once more.

By the beginning of the seventeenth century, however, most of the area had been enclosed and drained, the Rother estuary now being restricted to a comparatively narrow channel running from Appledore to the open sea near Rye with a subsidiary inlet, the Wainway Channel, flowing into it from the east (fig. 25).

The topography of the Marsh at this period is summed up with some accuracy on Poker's map of 1617. Thereafter the final reclamation of the Rother creek, the advance of Dungeness, the building of the beach across the Old Romney estuary and the reclamation of the land behind, and the erosion of the Dymchurch beach and its strengthening with the Dymchurch Wall, are all events which can be traced with precision on more recent maps. The statistics published by Lewis and Balchin (1940, 268–9) which show that the average annual advance of Dungeness (based on figures for the period 1878–1938) is between 3–4 yards (2.7–3.6 m) a year, are a reminder that the evolution of Romney Marsh is far from complete.

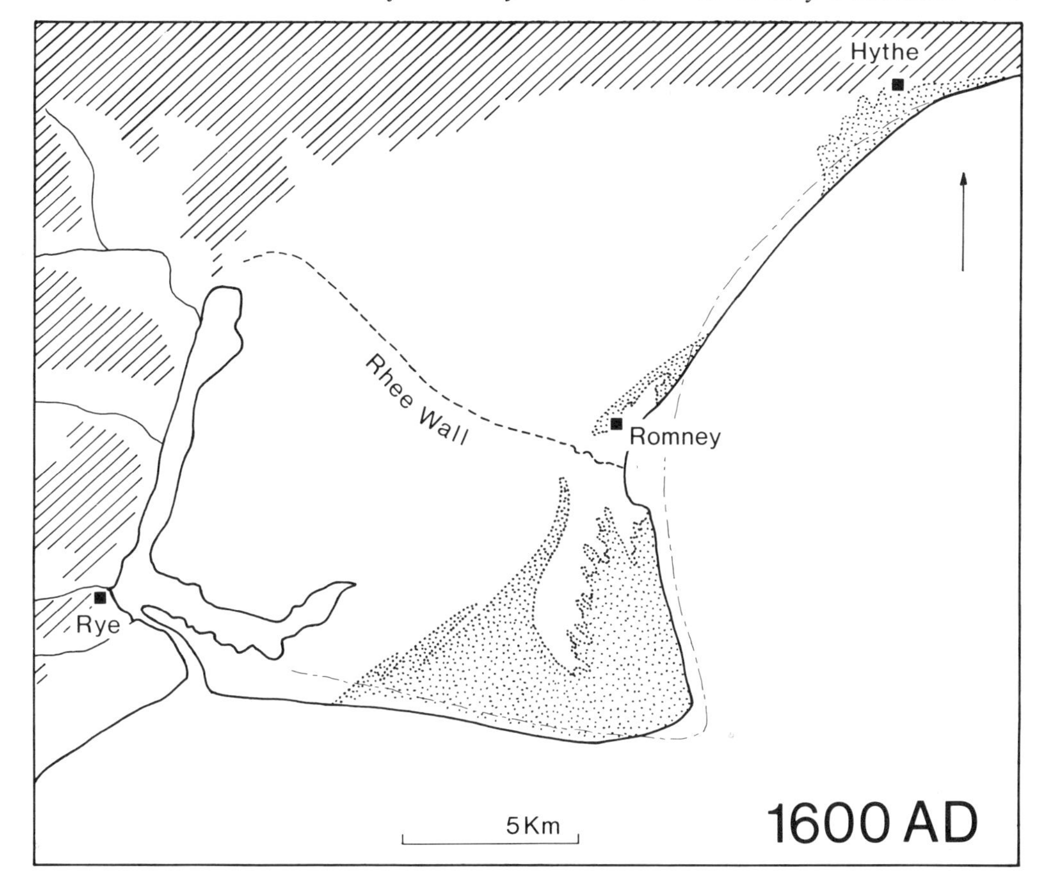


Fig. 25. The Romney Marsh region *c.* A.D. 1600.

## *Concluding Remarks*

Sufficient will have been said, in the all too brief outline given above, to show that it is now possible, largely as the result of the work of the Soil Surveys of England and Wales, to give a tolerably cohesive picture of the development of Romney Marsh. Details have been omitted and areas of obscurity passed over but after almost three centuries of debate the main stages in the process of landscape evolution can at last be presented with some assurance.

The most striking point to emerge from this preliminary survey of the evidence is the immense value of the Marsh as a research resource. In spite of the surface devastation wrought by the army and by the holiday industry and the destruction of large areas of old inland beaches by gravel working, the raw data still stratified in the marsh deposits represents a unique record of landscape development from the Neolithic period to the present day.

The survey focuses on several areas deserving further study:

(a) The buried Neolithic and Bronze Age forest landscape provides a rare opportunity for studying both the settlement of the period and the effects of man on his forest environment.
(b) In the early Roman period the Marsh is likely to provide evidence of specialised settlements and economies in increasingly stressed situations.
(c) In the middle to late Saxon period, the remarkably rich documentary and place name evidence together with topographical studies and, potentially, excavation would allow the colonisation of a virtually bare landscape to be examined in detail.
(d) The rapidly changing medieval landscape provides an ideal background for the study of medieval engineering as well as the socio-economic history of marshland communities.

Simply listing a few of the outstanding research areas is sufficient to underline the considerable potential of the Marsh for environmental, archaeological, and historical studies. The work of the Soil Survey staff, in clearing away many of the misconceptions and uncertainties which have bedevilled the subject in the past and in providing a thorough cartographic background, has initiated a new era in the study of the evolution of Romney Marsh.

## BIBLIOGRAPHY

Akeroyd, A. V. 1972. Archaeological and historical evidence for subsidence in southern Britain. *Phil. Trans. R. Soc. Lond.*, A. **272,** 151–71.
Cunliffe, B. forthcoming. Excavation at the Roman Fort at Lympne, Kent, 1976–8. *Britannia.*
Dowker, G. 1897. On Romney Marsh. *Proc. Geol. Assoc.* **15,** 211–23.
Drew, F. 1864. *The Geology of the Country between Folkstone and Rye including the whole of Romney Marsh* (Mem. Geol. Surv.: London).
Dunkin, A. J. (ed.) 1844. The late discoveries at Dymchurch. *Report of Proc. Brit. Arch. Assn. Canterbury,* 115–21.
Elliott, J. 1847. Account of the Dymchurch Wall, which forms the sea defences of Romney Marsh. *Min. Proc. Instn. Civ. Engrs.* **6,** 466–84.
—— 1852. Notes on the original plan of the castrum at Lymne, and on the past and present state of the Romney Marshes. In *Report on Excavation made on the Site of the Roman Castrum at Lymne, in Kent, in 1850.* by C. R. Smith. London.
—— 1862. Reported in Lewin (1862).
Gilbert, C. J. 1933. The evolution of Romney Marsh. *Arch. Cant.* **45,** 246–72.
Green, R. D. 1968. *Soils of Romney Marsh.* Harpenden.
—— and Askew, G. P. 1958a. Kent. *Rep. Soil Surv. Gt. Br.* No. 9, 27–30.
—— and —— 1958b. Kent. *Rep. Soil Surv. Gt. Br.* No. 10, 21–5.
—— and —— 1959. Kent. *Rep. Soil Surv. Gt. Br.* No. 11, 22–7.
—— and —— 1960. Kent. *Rep. Soil Surv. Gt. Br.* No. 12, 35.
Holloway, W. 1849. *The History of Romney Marsh.* London.
Holmes, T. R. E. 1907. *Ancient Britain and the Invasion of Julius Caesar.* Oxford, Clarendon Press.
Homan, W. M. 1938. The marshes between Hythe and Pett. *Sussex Arch. Coll.* **79,** 199–223.
Isaacson, S. 1846. Discovery of Roman urns and other ancient remains, at Dymchurch in Romney Marsh. *Archaeologia,* **31,** 487–8.
Jones, I. 1953. Roman remains on Lydd Rype. *Arch. Cant.* **66,** 160–1.
Lewin, T. 1862. *The Invasion of Britain by Julius Caesar with Replies to the Astronomer-Royal and of the Late Camden Professor of Ancient History at Oxford.* London.

Lewis, W. V. 1932. The formation of Dungeness foreland. *Geog. Journ.*, **80,** 309–24.
——and Balchin, W. G. V. 1940. Past sea-levels at Dungeness. *Geog. Journ.*, **96,** 258–85.
Milner, H. B. and Bull, A. J. 1925. Excursions to Eastbourne–Hastings. *Proc. Geol. Assoc.*, **36,** 317–20.
Rendel, W. V. 1962. Changes in the course of the Rother. *Arch. Cant.*, **77,** 63–76.
Smart, J. G. O. 1964. In National Physical Laboratory radiocarbon measurements II. *Radiocarbon*, **6,** 25–30.
—— 1966a. In *Geology of the Country around Canterbury and Folkestone*. (Mem. Geol. Surv.: London).
—— 1966b. In *Geology of the Country around Tenterden.* (Mem. Geol. Surv.: London).
Smith, C. R. 1847. On Roman pottery discovered in Kent. *Journ. Brit. Arch. Assn.*, **2,** 132–60.
—— 1850. *The Antiquities of Richborough, Reculver and Lymne in Kent etc.* London.
Steers, J. A. 1964. *The Coastline of England and Wales.* Cambridge, University Press.
Teichman Derville, M. 1936. *The Level and Liberty of Romney Marsh, Kent.* Ashford.
Tooley, M. J. 1974. Sea-level changes during the last 9000 years in north-west England. *Geog. Journ.*, **140,** 18–42.
Wallenberg, J. K. 1931. *Kentish Place-names: a Topographical and Etymological Study of the Place-name Material in Kentish Charters dated before the Conquest.* Uppsala.
Ward, G. 1931a. Saxon Lydd. *Arch. Cant.*, **43,** 29–37.
—— 1931b. Sand Tunes Boc. *Arch. Cant.*, **43,** 39–47.
—— 1933a. The Saxon charters of Burmarsh. *Arch. Cant.*, **45,** 133–41.
—— 1933b. The river Limen at Ruckinge. *Arch. Cant.*, **45,** 129–32.
—— 1936. The Wilmington charter of A.D. 700. *Arch. Cant.*, **48,** 11–28.
—— 1940. Discussion in Past sea-levels at Dungeness, by W. V. Lewis and W. G. V. Balchin. *Geog. Journ.* **96,** 281–2.

# Iron Age and Roman Coasts around the Wash

Brian Simmons, F.S.A.

*Introduction*

Before examining coastal changes around the Wash two points should be made relating to this paper. First of all the terms 'coasts' and 'coastlines' are, perhaps, misnomers when applied to the Lincolnshire Fens as they imply fixed high water marks with the sea washing regularly up to a shoreline. A better description would be a sea-affected marsh, a scene typical of the present day Lincolnshire coast. Secondly, research into the subject of coastal changes in this part of Lincolnshire is by no means complete; on the contrary, work is still progressing and as more information becomes available, then ideas on the precise detail of inlets, bays, creek patterns, islands and so on will, undoubtedly, alter and amend thinking on the everchanging sea and consequent land settlement patterns.

The soils of the Lincolnshire fens are of two different and general types, peat and silt (fig. 26). To the north there are the Witham peat fens and peat occurs again from Bourne southwards, swinging through Cambridgeshire and into Norfolk in a crescent shape. Between the peats and stretching from the western fen edge to the sea are the silt fens, a completely different type of soil from peat. It is not intended to give a definitive account of the silts in this paper except to say that they vary in nature from clays, and clayey gravels, through to alluvium (O.S. Geol. 1912 and 1954.)

To the west, away from the fens, the land rises to about 200 ft. (61 m) at its maximum height. The land here is predominantly limestone and clay. From this maximum height of 200 ft. (61 m) and moving eastwards through the silt fens the land slowly drops to a minimum height above sea level of about 5 ft. (1.5 m) or lower in the centre of the fens and then gradually rises to about 20 ft. (6.1 m) on the coast. The fens should be seen as a natural trough and this effect will be taken into consideration when ancient coastlines are examined.

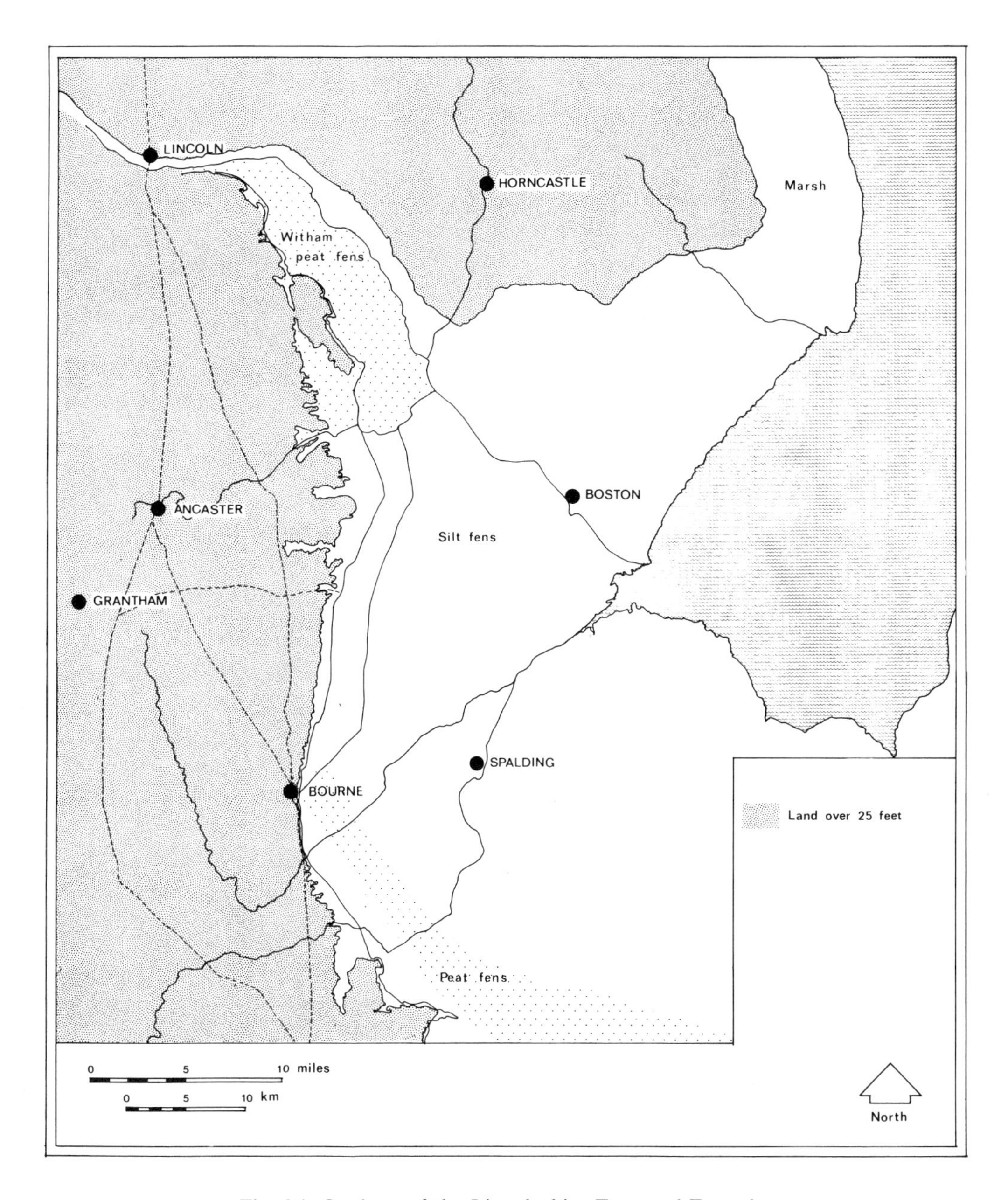


Fig. 26. Geology of the Lincolnshire Fens and Fen edge.

## Methods of Research

The techniques used in the research were simple and straightforward. First of all, aerial photographs, obtained from many sources, were used and all necessary information drawn onto O.S. maps. The information was not only archaeological in character; the recording of ancient creeks, for example, also proved to be vital.

It was fortunate that, during the course of the study, a good working relationship was built up with the main drainage boards in the fens, the Anglian Water Authority and the Black Sluice Internal Drainage Board. Not only were they prepared to give prior notice of their earthmoving programmes, but they also made freely available their records and supplied between them some 7,000 land levels which were crucial in the understanding of the present terrain. These land levels were plotted onto overall maps of the area and contours were made at various intervals; this work will be dealt with later.

Another source of help came from the Soil Survey of England and Wales. It is true to say that without their assistance the research could not have progressed as it has done. Although there is only one publication of soils within the area, the Woodhall map (Robson *et al.* 1974), the surveyors allowed the use of unpublished material. They also spent many hours in the field checking on soil types, giving advice on the interpretation of aerial photographs, and lending equipment to further the interpretation of the soils.

Other help came from the engineers at the Hydraulics Establishment at Wallingford who have recently completed their work on the Wash Barrage, and were prepared to release some of their findings.[1] Some palaeobotanical results were also obtained, as well as snail types, C 14 dates, and the Anglian Water Authority gave permission for its map of the 1953 flooding to be used.

The field-walking programme included the recovery of *all* pottery and other artifacts from the surface of fields. At the same time any change in soil colour, particularly when associated with other information such as undulations in hedge lines or on the surface of the fields, was carefully recorded. Frequently, commercial earthmoving, especially during the making of large drains, was revealing.

On some sites, particularly Roman ones in the Fens, the archaeological fieldwork was developed beyond the usual practice of walking up and down the field making notes, taking photographs and collecting pottery. Sometimes it was possible to use five methods of studying the same site; aerial photographs,[2] contour surveys, soil results, archaeological remains on the surface and, occasionally, palaeobotanical evidence. In Hacconby Fen, for example, a grass field revealed, by aerial photography (Hallam S. J., in Phillips 1970, pl. XIa), part of a much larger Roman settlement spreading over at least 100 acres (40.5 ha.). This has been interpreted as a large, loose settlement with small enclosures, linked by droves (Hallam S. J., in Phillips 1970, 263). A second aerial photograph (Hallam S. J., in Phillips 1970, pl. XIb) taken a short time after the field had been first ploughed in 1956 seemed to confirm this view. The fact that ploughing helps to reveal more detail of occupation than when the land was under grass is, perhaps, a doubtful bonus. A detailed drawing of the marks seen from the air was produced.

Although the Hacconby field has been ploughed every year since 1956, and chisel-ploughed on at least two occasions, the ditches and areas of soil discoloration can still be seen at ground level. It was possible to survey these features onto another plan of the field. At the same time, a detailed distribution of the pottery found on the surface of the field was made and yet another plan made of this. The last piece of research done at Hacconby was to examine the soils and the results are clear;[3] the drove ways are more likely to be roddons or tidal creeks. The dark, peaty soils associated with the enclosures, seen on the aerial photographs and on the ground with the pottery scatters, are the remains of human occupation.

A final composite plan (fig. 27) of the field bringing together all the information now gathered suggests a different interpretation for Hacconby Fen. In the south-east corner of the field there is a Roman salt-making area, with the roddons supplying sea water to the salterns. The enclosures with their associated black soil and pottery are on raised ground and within each enclosure it is possible to define the huts, even after 20 years or so of intensive arable agriculture (Simmons, in Schadla-Hall and Hinchcliffe). It is noticeable that on this type of site in the fenlands pottery does not move around the surface of the field as much as might be thought; equally the soil discolorations remain visible over long periods.

On the eastern side of the Hacconby site are two faint lines representing part of an ancient watercourse known in the early Middle Ages as the *Midfendic* (Hallam S. J., in Phillips, 1970. pl. xɪb, and Hallam H. E. 1965, 64–5). More will be said of this watercourse shortly, but its relationship with a second watercourse, the so-called Bourne-Morton Canal is significant (Hallam S. J., in Phillips 1970, pl. xɪb and 254–5). There can be little doubt that both come together close to the salt-making area in Hacconby. In summary and from all this information it can be demonstrated that the Roman site in Hacconby was affected by sea water. There seems little chance of intensive arable agriculture having been possible in this type of environment as suggested by some writers (Richmond 1963, 129–30). The earliest date of the surface pottery recovered from Hacconby is the first half of the second century A.D. and is in accordance with 120 or so other fields examined in these fens.[4]

The distribution of known Roman finds from the Lincolnshire fens and fen edge (fig. 28) has been compiled from information taken from published sources (Hallam S. J., in Phillips 1970, 175–331; *EMAB* 1959–77; *LAASRP* 1958–64; *LHA* 1963–77; Creasey 1825; Marrat 1816; Phillips 1934; Swinnerton 1932; Trollope 1872; Nenquin 1961), museum records,[5] and a field walking programme. Each triangle on the map represents at least 100 sherds, usually when associated with other evidence: changes in soil colour, humps and bumps in the fields and hedge lines, marks on aerial photographs and so on. More frequently, however, the sites represent large areas of settlement, in one case of more than 200 acres (80.9 ha.) (Sempringham Fen) and in many other cases of more than 50 acres (20.2 ha.) each. It was also noted during this work that the type of site seen falls into the native settlement pattern; there are no known villas or towns within the Lincolnshire fens.

The field work undertaken for this research also demonstrated other facts which are important when the question of coastlines is raised. First of all there are few tiles, and no building stone to be seen from any of the surface collections (Simmons 1975a, 20). Of the large quantity of pottery retrieved from these fens, only a handful of sherds are earlier than the first half of the second century A.D. The aerial photographs show few field systems which can be related to the vast farming enterprises usually associated with the fenlands (Salway, in Phillips 1970, 20).

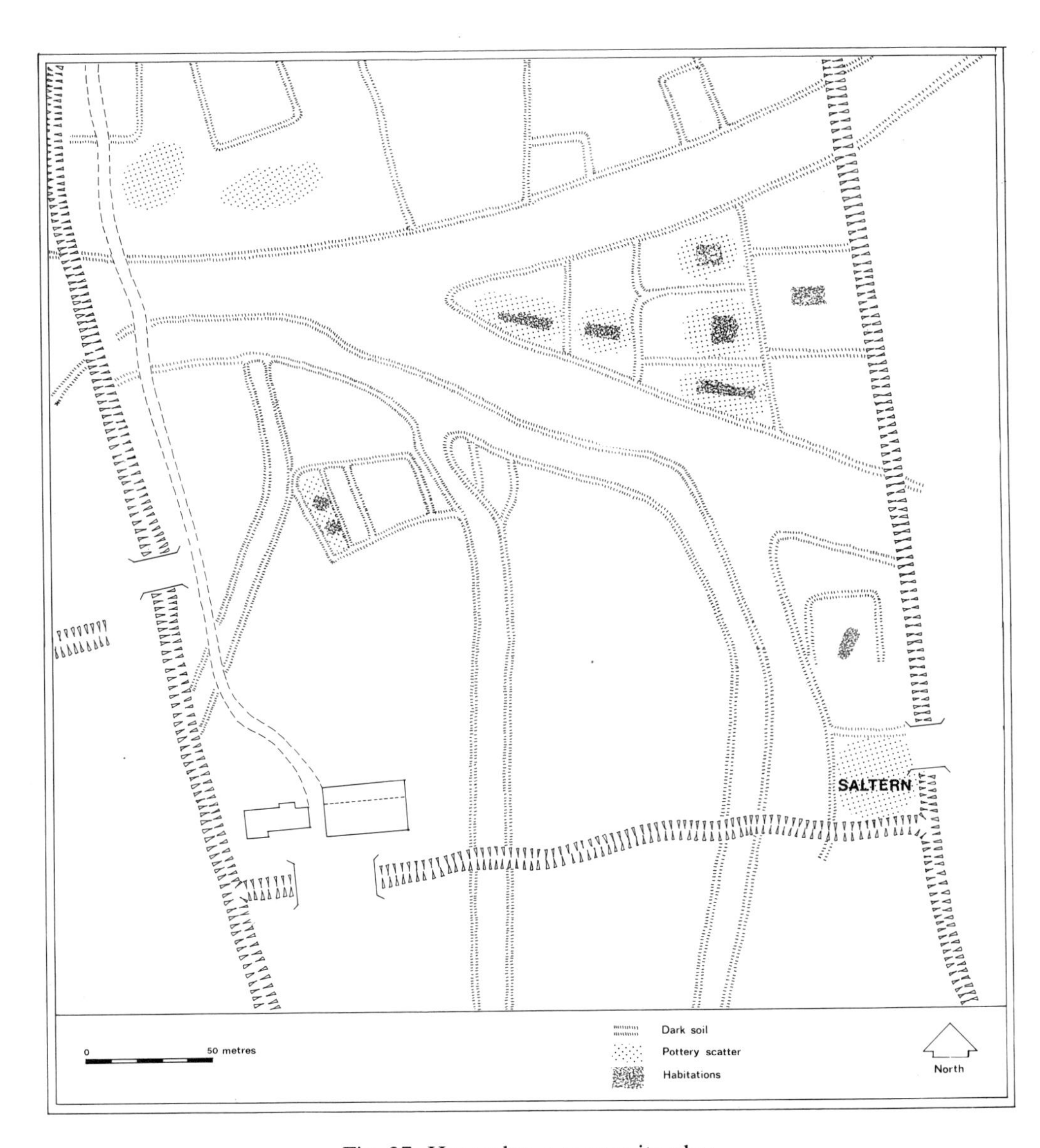


Fig. 27. Hacconby: a composite plan.

Lastly, there are no known finds from the peat fens, and very few, if any, aerial photographs give any hint of the settlement patterns related to this industry or any other type of Roman occupation. The impression gained of settlement is one of a low standard of living (small insubstantial huts with few material possessions) in the silt fens, but nothing in the peat.

The Iron Age picture (fig. 28) is unlike that of the Roman period. There are some Iron Age finds in the fens and where these sites occur they are generally on the higher land on the eastern seaward fen margins, for instance, in Wrangle and Holbeach. Other major sites are inland along the line of the Car Dyke where some of the occupation areas are relatively extensive, up to 30 acres (12.1 ha). It is of interest, in this study, to compare the early Saxon settlement with the Iron Age. The early Saxon finds are to the west of the Car Dyke, or immediately along its course, but by the seventh or eighth centuries it appears that some land along the higher eastern ridge, Fleet, Gedney, Fishtoft, Algarkirk, for example, had become available, for one reason or another, for habitation.[6] The environment during the three or four centuries after the withdrawal of Roman administration seems to compare with the climatic and marine situation before the Roman occupation of the silt fens.

A broader view can now be taken of Hacconby and other Iron Age and Roman sites. From the original collections of pottery it is possible to select distinctive types of artifact. Of these salt-making briquetage is perhaps the most revealing in many ways. Most of the briquetage comes from surface collections but one Iron Age saltern site in Helpringham Fen has been partly excavated and the mounds, ditches and hearths from the excavation represent an industry which covers an area of at least 30 acres (12.1 ha) (Simmons 1975c). Seeds from the excavation, particularly of the species seablite,[7] confirm that the influence of the sea came within a few yards of the salt-making process.[8] There are many more sites of a similar nature along the line of the Car Dyke in the silt fens (fig. 29). Other Iron Age salterns occur in Holbeach and Wrangle, and it is assumed that by about 200 bc the industry was well established, especially along the line of the Car Dyke (Simmons, in de Brisay, 1975, 34).

To the east of the Car Dyke, between 2 and 3 miles away (3.3–4.9 km), is a second line of salterns, but this time of a Roman date; the surface evidence seems to be quite positive on this count. The arrangement of Roman salterns takes up another line (fig. 29), this time along a watercourse, the *Midfendic*. Other groups of Roman salterns are known elsewhere in the Lincolnshire Fens, notably near Holbeach and Whaplode, Wrangle, Swineshead and Burgh-le-Marsh. It should be reiterated that there is no known salt making, either Iron Age or Roman, in the peat fens.

The methods of elucidating the evidence have been put forward and the processing of the information has been mentioned, albeit briefly. Before moving on to discuss a composite picture of the work so far and the suggested coastlines around the Wash, an examination should be made of some earlier work which has a bearing on coastlines here. The original intention of the research (which commenced some ten years ago) in the Lincolnshire Fens was to examine the Car Dyke in detail.[9] The O.S. map of Roman Britain shows the supposed course of the Car Dyke from Cambridgeshire to Lincoln. Most books which refer to this monument

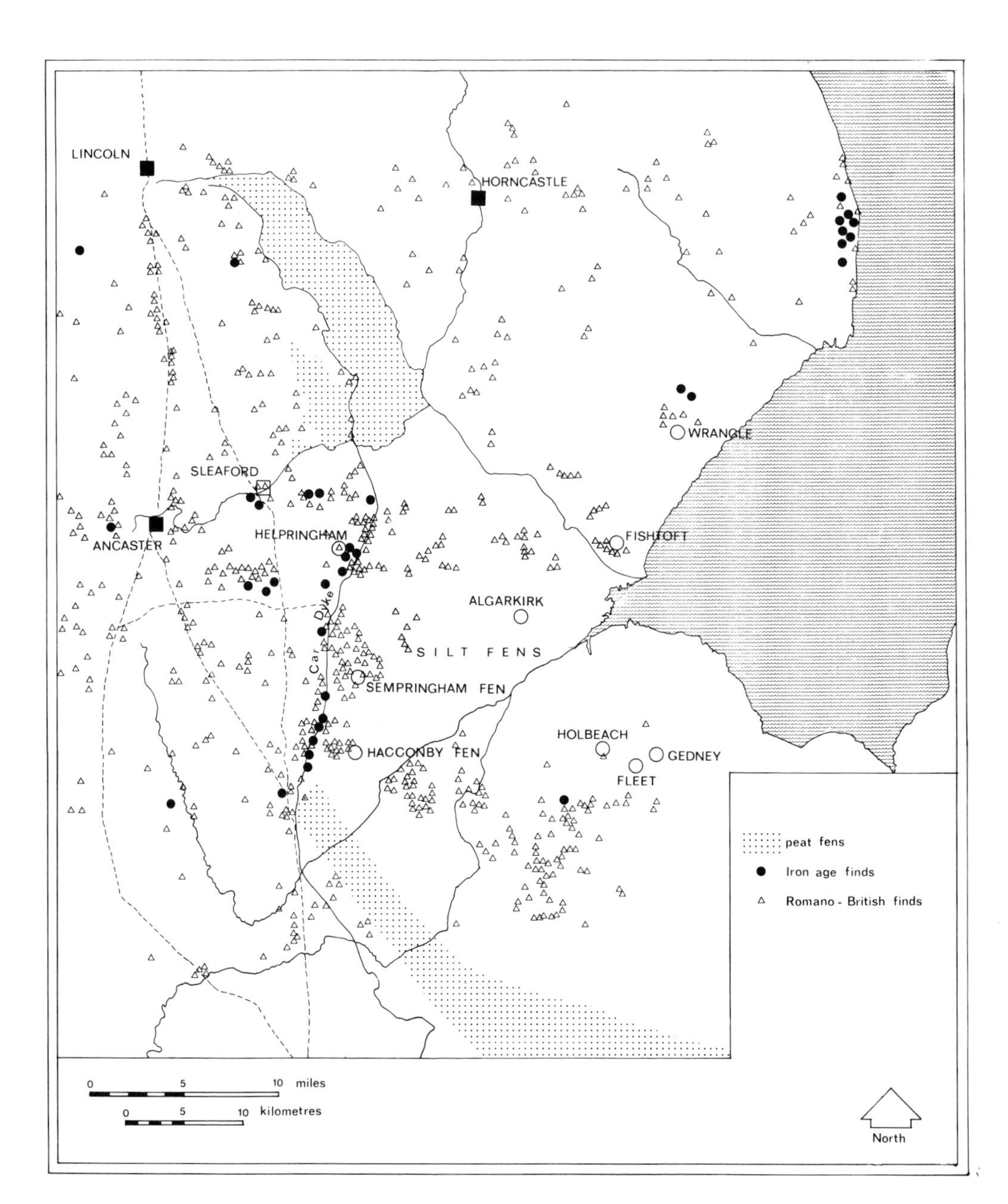


Fig. 28. Distribution of Iron Age and Roman finds.

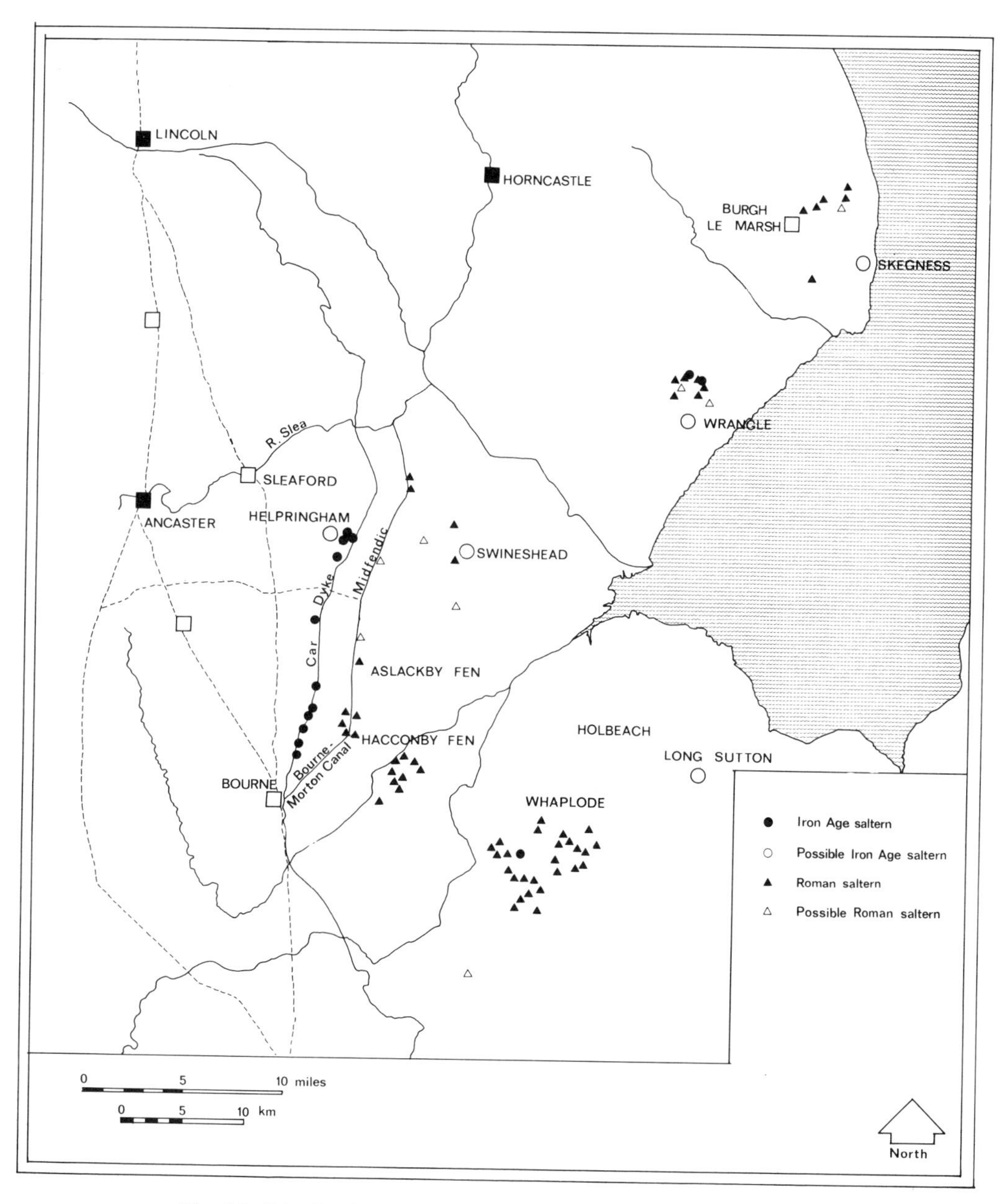


Fig. 29. Distribution of Iron Age and Roman salt-making sites

regard it as a Roman canal used for transporting the considerable agricultural produce, mainly grain, of the fenlands to the major markets in the north, notably Lincoln, and possibly York. In fact, the Car Dyke should be seen in conjunction with three other major watercourses, the *Midfendic*, the Bourne-Morton Canal and the River Slea, and the many west–east minor natural streams. It is not the intention of this paper to go into the whys and wherefores of the Car Dyke but merely to state that it may well be to do with a Roman drainage system in the silt fens and not with long-distance transport. A large corpus of evidence relating to the Lincolnshire Car Dyke has now been collected which shows that the silt fen Car Dyke may have been used as part of a larger drainage scheme (Simmons 1975a, 83–128; Simmons 1975b; Simmons 1979). There is nothing to suggest otherwise, unless the watercourse was also a political or economic boundary.

It is, however, the parallel channel to the Car Dyke, the *Midfendic*, which should now be examined with regard to the Roman coastline. A drainage system of any sort in the conditions which appertain to the fenlands must satisfy three conditions: one, to withhold the highland water, which the Car Dyke does; the second, to absorb the surface and rain water, which the minor, west-east streams achieve; and the third is to safeguard the new land from inundations from the sea. This latter requirement generally takes the form of a sea-bank on the seaward side of a catchwater drain. A catchwater drain has three functions in itself: (i) to act as a surface water drain; (ii) to fulfill the purpose of a second reservoir, a safety valve for any water which cannot be stored in the highland drain and has to be moved through the new lands; (iii) to hold any sea-water which spills over the sea-bank during abnormal tides. If such a catchwater drain exists, incorporated within a drainage scheme and associated with a sea-bank, then a coast may be inferred.

By chance, in 1975, the Black Sluice Internal Drainage Board widened and deepened a drain in Aslackby Fen and very close to the *Midfendic*, in an area where a large coverage of Roman saltern debris, 40 acres (16.2 ha) or more, had already been noted on the surface of fields during field walking. The drainage engineers gave permission to clean that part of the section which was of archaeological interest (fig. 30) (Simmons 1975d). The new drain runs west–east and cuts through three pits or ditches which are at right angles, more or less, to the modern channel. About 200 yd. (182 m.) to the west is the old line of the *Midfendic*. Closest to the *Midfendic* was a pit filled with green clay. The next feature was peat and silt filled, with abraded briquetage present. A further similar feature occurred a few yards to the east and up to a large silt bank. Included in the silt bank were layers of briquetage in the very clean silt. It is important to compare this briquetage with that occurring in the pits or ditches and in the dirty silty layers which were on either side of the silt bank. In every instance, with the exception of the silt bank, the briquetage had the appearance of being water-borne, whereas that briquetage sealed in the bank had sharp, crisp breaks. It is not difficult to see the erosion of a larger sea-bank and the ensuing dirty silt with briquetage, abraded in the process, deposited on the seaward side, as depicted in the drawing (fig. 30).

The salt-making industry here is exactly where it would be expected, behind the protection of a sea-bank or breakwater, but on the salt-water side of the fresh-water *Midfendic*. A similar situation can be seen in Helpringham Fen where the *Midfendic* can be clearly observed, now as a partly filled-in channel. Immediately to the east of it is a Roman salt-making site and beyond this the protecting bank or breakwater. Another possible bank has been recorded in Hacconby (Hallam, S. J., in Phillips 1970, 265). A further glance at the distribution of salt-making sites (fig. 29) in the Lincolnshire silt fens should put the relationship of *Midfendic*, sea-banks and salt making into perspective.

## *Coastlines*

From the evidence thus far presented, a picture of coastlines at two dates can be deduced, 200 bc and the second century A.D. These two dates are somewhat arbitrary and are based, on the one hand, on the average date of the Iron Age pottery recovered from the many sites discovered along the landward fen edge (together with the radiocarbon date mentioned above), and on the other hand the earliest date of the pottery from the many Roman sites in the fens, particularly along the *Midfendic*.

The outline of the Roman coast has been built up in much the same way that the Hacconby Fen settlement was examined, although in broader detail. In Hacconby it can be demonstrated that the tidal roddons are man-altered and that the hut platforms are neatly aligned onto the roddons. Furthermore, there is a saltern in Hacconby. It is, therefore, reasonable to suppose that the effect of the sea was present in Hacconby during the Roman period. A similar situation can be seen over and over again in the silt fens. When five different overlay maps are produced for five different aspects of the fens, archaeological distribution, aerial photographic information, soils including ancient creeks, contours produced from land levels, and finally a map including palaeobotany, the Anglian Water Authority's 1953 flood penetration, and so on, the result is a much different coastline from the modern one (fig. 31).

For instance, the general distribution map of Roman finds (fig. 28) appears to give a blanket occupation for the fens. In fact, all the Roman occupation, almost without exception, is contained within a contour of 10 ft. (3 m) above sea-level. The islands shown on this coastline map are all above 10 feet. It must be more than coincidence that nearly all the salt-making sites on these islands are on the leeward side of each island, not only affording protection for the salt-makers from the harshness of the North Sea, but also increasing the salt content of the sea-water (de Brisay 1972–4, 22–23). Creeks on the back of the islands are also more likely to be associated with tidal ponds which would act as evaporating basins increasing the salinity of the tidal waters.

A study of the soils in various areas is also rewarding. The island where modern Sibsey is situated can be shown to be ancient from a study of its soils.[10] The island of Sibsey is also above the 10 ft. (3 m) contour and, again, has within its area Roman settlement sites. The planning of the soils around Sibsey from aerial photographs taken especially for the purpose duplicates the island which was plotted independently using the other information.

In other parts of the fenlands a similar pattern can be produced over and over again, but it would be impractical to discuss this detail here. Nevertheless, it may be worthwhile to mention the new outfall of the river Witham, as this will have some bearing on what is said soon. The ancient Witham can be seen (together with the rivers Slea, Bain and, perhaps, the northern part of the Car Dyke) discharging

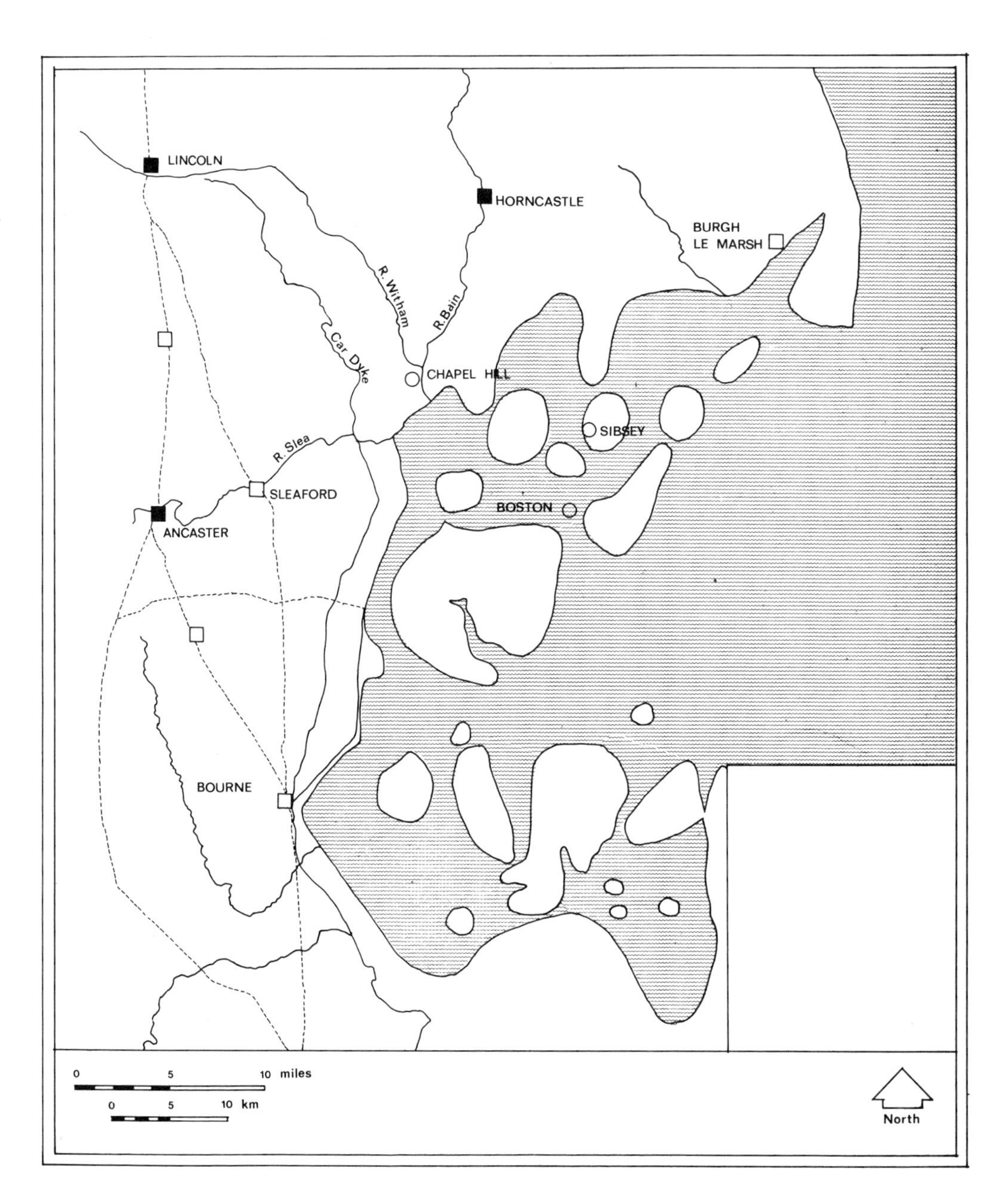


Fig. 31. Roman coast around the Wash *c.* A.D. 200.

into the sea at Chapel Hill (fig. 31). This delta was much further inland than the modern estuary, which is near Boston. It is interesting to note that the soil immediately to the south of the ancient delta is made up over a wide area of riverine and estuarine clays.[11]

The Iron Age coast (fig. 32) takes a different form from the Roman one. Salt-making sites along the Car Dyke have been mentioned and these, together with the plant remains in the Helpringham excavation, are suggestive of a coast-line here. Other Iron Age finds in the fens are sparse, and are only found on the higher coastal strip. Again, the use of land levels to build up a contour map demonstrates that all the Iron Age finds are on land at 15 ft. (4.6m) or higher. In other words, there is an apparent drop in the sea-level of 5 ft. (1.5 m) during the 400 years or so of the periods discussed in this paper. This marine regression, Iron Age to Roman, of 5 ft. (1.5 m) has been noted by other workers. One of the most recent of these researchers is Dr M. J. Tooley, who made a study of Morecambe Bay. His published graph gives a fairly similar situation for the period in question (Tooley 1974, 32).

It would be foolish to suggest that this is the final word on Iron Age and Roman coasts around the Wash. There are problems and pitfalls to overcome and some of the questions may never be answered. Land tilt and erosion present almost insuperable difficulties. What can be said of the loss of old Skegness, reputed to be drowned under the sea 3–4 miles (4.9–6.5 km) east of modern Skegness (fig. 29)? (Leland quoted in Phillips 1934, 134). Is there a Roman fort here as some people have suggested? (Johnson 1976, 20, 123; Whitwell 1970, 52). Further around the coast in the Long Sutton area (fig. 29) there is a great depth of silt and there have been reports of Roman remains buried at a depth of 4 ft. (1.2 m) or more (Hallam, S. J., in Phillips 1970, 311–2). Is there post-Roman silting masking large areas of settlement which are not now discernible? It is difficult to see how these questions can be easily solved by traditional archaeological methods.

There is, however, one problem which can be overcome and this is the question of land shrinkage, or compaction, within the silt fens. Doubts have been cast on the value of using modern land levels and applying these to Roman or Iron Age landscapes. It is possible to look at one particular period of 60 years or so to establish whether compaction has taken place or not. From before the First World War until well after the Second the Lincolnshire fens underwent a considerable agricultural change from almost total pasture to almost total arable, and from horse-drawn ploughs to tractor-drawn ploughs. Both these factors are important. On the one hand if the land is going to shrink (or expand) major changes in land usage are going to be contributory factors; rubber-tyred tractors, for example, are known to consolidate the soil beneath the plough horizon and cause a false water table (Spoor, in Schadla-Hall and Hinchcliffe); to plough, totally, vast areas of grass can cause a drying-out of the soil, or even soil erosion. The situation can be tested by comparing similar sections made through the fens by the Drainage Boards in 1913 and again in 1970. It can be seen from this comparison that there are only a few inches difference either way, and the overall effect is one of no change.[12]

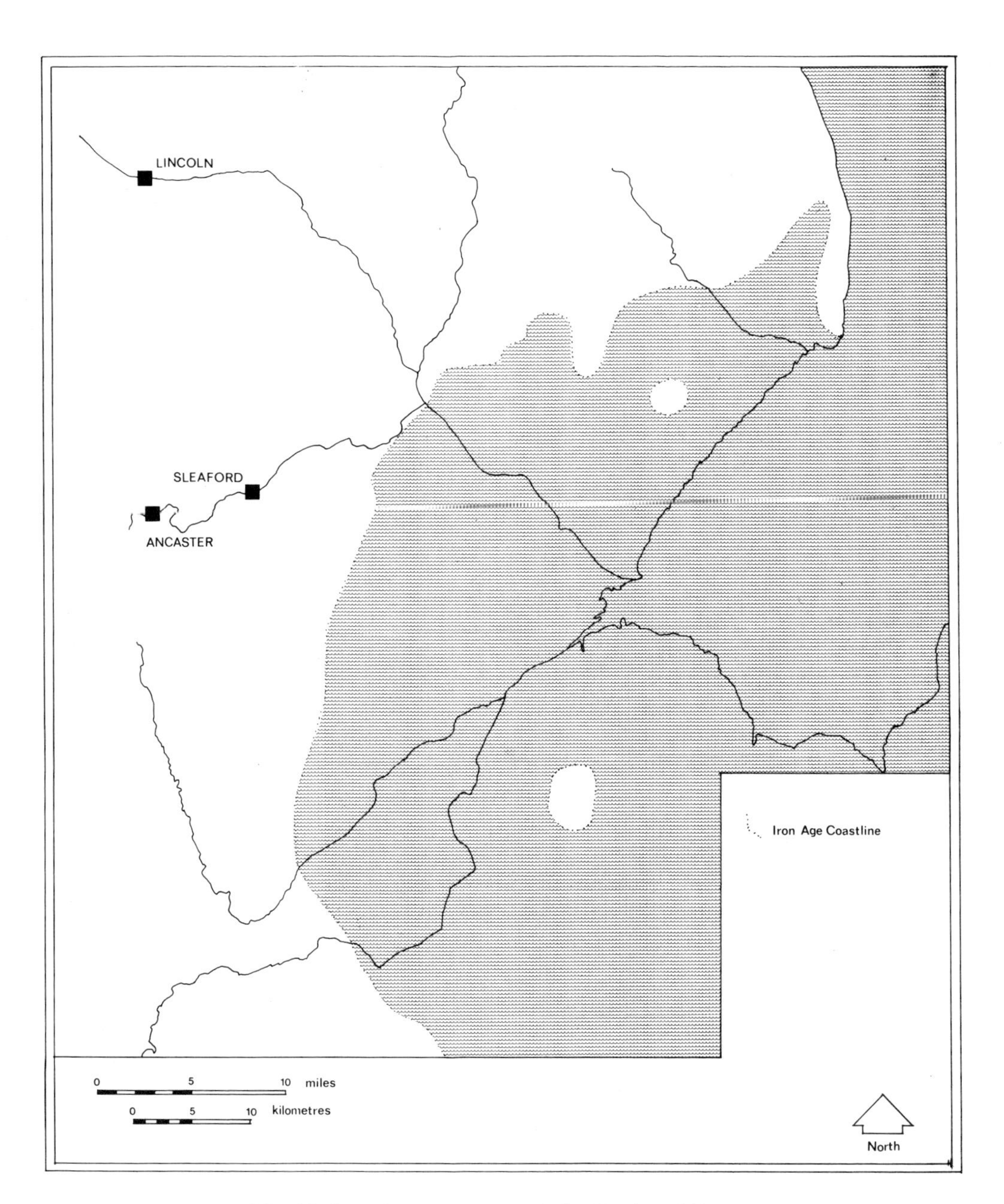


Fig. 32. Iron Age coast around the Wash *c.* 200 B.C.

### *Future Programme of Work*

It is obviously necessary for more work to be done on the question of coastlines around the Wash and a future programme of work can be instituted. A better understanding of the soils is needed and for this reason sections made by Drainage Authorities from time to time can be examined in detail. Already a long section has been studied in Horbling Fen (Simons and Chowne 1975), and the layers examined by soil surveyors. In particular the thin buried peat is of importance. A research project has been set up throughout the Lincolnshire fenlands in order to obtain as many samples from as wide an area as possible to see if there is any correlation between the buried peats. Carbon dates are of vital importance in this work,[13] as well as the study of plant remains. When there is a better understanding of the buried peats, then other ancient coastal changes may be evaluated. Excavations on selected saltern sites, too, might be revealing.

The Bronze Age is the next obvious period to tackle, and although only the Roman and Iron Age results of the field-walking project have been mentioned, recent Bronze Age finds are surprising. These range from new barrow groups, some with 15 or more barrows per group, areas of Bronze Age pottery on the surface along the silt fen edge (for instance, 60 acres (24.3 ha) in Heckington and several miles of ditches and settlement areas in Billingbrough) (Chowne 1978), and possible Bronze Age industrial sites, especially salt-making. Much more work needs to be done, however, before any coastline conclusions for this, and earlier periods can be achieved.

In other directions, particularly Roman, new thoughts and ideas come to mind. The usual tribal territories shown around the Wash (Cunliffe 1974, 87) apportion the southern part of the modern coastline to the Iceni and the northern part to the Coritani, with the Catuvellauni well inland (fig. 33). Using the new coastline map the Catuvellauni could well have its own coast in the central part of the Wash. More than this, the Catuvellauni could then have at least one off-shore island with a large group of Roman salterns (together with one Iron Age saltern) in the Holbeach/Whaplode area. Is this area the salt works of the Catuvellauni to which Ptolemy refers (Rivet 1964, 131–2)?

Roman arable agriculture in the Lincolnshire fens has been said to be all-important, but is this thought tenable any more? There is little evidence to support the view. Aerial photographs show a singular lack of field systems, and large grain-growing farms in a salt-affected landscape are not easy to prove. Salt and crop growing are not happy bed-fellows. Should not consideration be given to other means of economy, possibly cattle-raising on a relatively small scale, fishing and oyster beds, basket making from reeds, and salt-making? Salt is necessary for tanning and preserving meat, and it could also be used for keeping oysters and other fish.

Some consideration may also be given to another aspect of Roman Britain as it affects the area of the land around the fenland basin. The late third- and fourth-century position of the various Roman towns which are closest to the Wash, Horncastle, Ancaster, possibly Burgh-le-Marsh, Sleaford and Bourne, all dominated to a greater or lesser extent by Lincoln, look vulnerable to a determined attack from North Sea raiders. It has been postulated that a Roman naval patrol from Brancaster to Skegness might have been the line of defence, but the entire

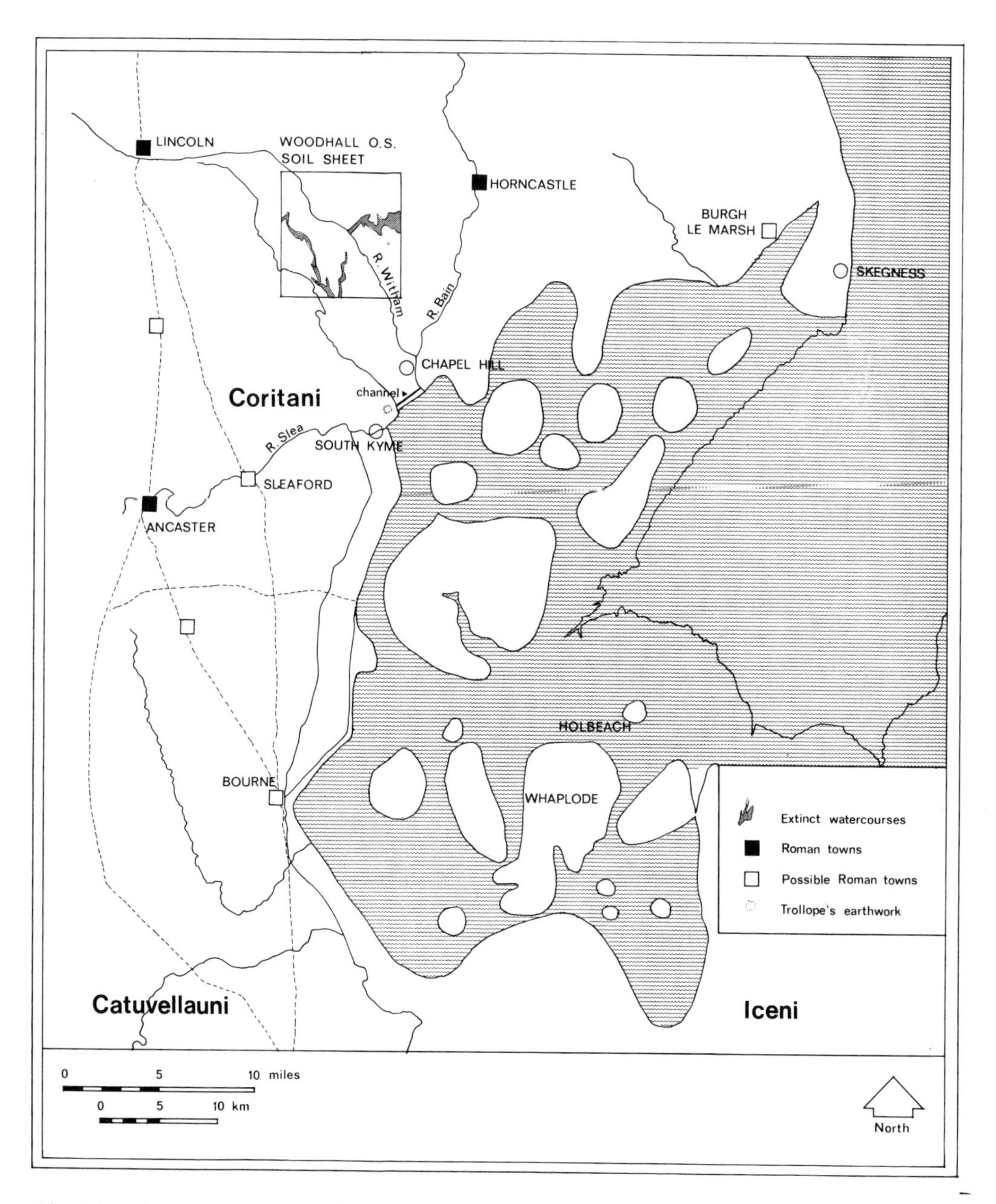

Fig. 33. Tribal territories in relationship to Roman coastline; Horncastle in relation to Roman coastline; extinct watercourses near Woodhall spa.

question of the defence of the Lincolnshire coast is open to speculation. Use of the new coastline might perhaps suggest another idea. Each of the three known Roman towns stands on a river: Lincoln on the Witham, Horncastle on the Bain and Ancaster on the Slea. These three rivers come together into a common estuary at Chapel Hill, and recently an extinct man-made channel has been discovered connecting the northern Car Dyke with the Witham (Simmons 1975a, 69–70). Furthermore, it is immediately on the landward side of the Roman coast in this part of the fens. That the watercourse is old can be ascertained by looking through the relevant records; there is no mention of it in post-medieval drainage accounts (*BSIDB*), nor is it likely to be medieval in construction as there was little activity in the area during the Middle Ages (Hallam H. E. 1965, 97; Coningsby appears to be exceptional in this context). Its most likely date is Roman, but if so, for what purpose? Perhaps the best answer lies in a late Roman defensive position; it can be argued that the most convenient place to defend the three towns of Lincoln, Ancaster and Horncastle would be in the neighbourhood of Kyme. A relatively small, mobile, force could have been kept to safeguard the approaches to the towns. It is interesting that in 1872 a small earthwork was recorded in Kyme and its plan published (Trollope 1872, 78). Could this be to do with late Roman coastal defence? Its position looks tantalisingly possible (fig. 33).

Lastly, we could turn our attention to the enigmatic position of Horncastle.[14] Horncastle has been thought of as part of the coastal defensive system (Todd 1973, 44), and it becomes an even stronger candidate in its relationship to the new Roman coastline. It is well positioned on a small tidal river, but there is a possibility that it was also on another tidal river. Work done by the Soil Survey of England and Wales on the area of the Ordnance Survey's sheet of Woodhall Spa (fig. 33) has not only indicated an extinct watercourse parallel to the Car Dyke (surely the Car Dyke here is natural?), but also another watercourse directed towards Horncastle and associated with the Witham (Robson *et al.* 1974, soil map). The surface soils are mainly estuarine clays; are they the remnants of a river once connecting Roman Horncastle with the Witham?

A research topic such as coastal change is fraught with danger; too much can be made of too little if care is not taken. The main purpose of this paper is to demonstrate that man's settlement is determined by natural forces which are almost beyond his control, although, in the Lincolnshire fenlands, the Romans were able to alter, to some extent, their environment. An equally important purpose of this paper is to try to stimulate further research into other archaeological aspects of the changing landscape around the Wash.

*Acknowledgements*

I am grateful to the following for their help and co-operation in this research: Mrs B. Kirkham, Mrs E. H. Rudkin, Miss R. H. Healey, Mrs C. E. Redmond, Mrs J. Cholmeley, Mrs G. Wilson, Dr H. Keeley, Mr J. Aram, Dr R. Evans and Mr D. Robson of the Soil Survey of England and Wales, Mr P. Chowne, Mr J. S. Wacher, Members of the Car Dyke Research Group, the Black Sluice Internal Drainage Board (particularly Mr F. E. Leafe and Mr J. R. Stokes), and the Anglian Water Authority (particularly Mr S. Cook).

Grants towards the cost of the research were obtained from the Society of Antiquaries of London, the Carnegie United Kingdom Trust, the University of Evansville, U.S.A., North Kesteven District Council and Lincolnshire County Council, and I should like to record my appreciation for their generosity.

## ABBREVIATIONS

| | |
|---|---|
| *BSIDB* | *Black Sluice Internal Drainage Board's Minute Books* |
| *EMAB* | *East Midlands Archaeological Bulletin*, 1959–1977 |
| *LAASRP* | *Lincolnshire Architectural & Archaeological Society Reports & Papers,* 1958–1964 |
| *LHA* | *Lincolnshire History & Archaeology,* 1963–1977 |
| O.S. Geol. 1912. | *Ordnance Geological Survey of England & Wales*, Drift Sheet 12 (1/4 inch to 1 mile), 2nd edn. |
| O.S. Geol. 1954. | *Ordnance Geological Survey of Great Britain (England & Wales),* Drift Sheet 143 (1 inch to 1 mile), revised 1967. |

## BIBLIOGRAPHY

Chowne, P. 1978. Billingborough Bronze Age settlement: an interim note. *Lincs. Hist. and Archaeol.,* **13,** 15–21.

Creasey, J. 1825. *History of New and Old Sleaford.* Sleaford.

Cunliffe, B. 1974. *Iron Age Communities in Britain.* London & Boston.

de Brisay, K. 1972–4. Excavation of the Red Hill at Osea Road, Maldon, Essex. *Colchester Archaeological Group Bulletin,* **15, 16** and **17** (reprint).

—— and Evans, K. A. (eds.). 1975. *Salt: the Study of an Ancient Industry*. (Colchester Archaeological Group) Colchester.

Hallam, H. E. 1965. *Settlement and Society.* Cambridge.

Johnson, S. 1976. *The Roman Forts of the Saxon Shore.* London.

Marrat, W. 1816. *History of Lincolnshire.* Vols. I & III. Boston.

Nenquin, J. 1961. *Salt, a Study in Economic Prehistory*. Bruges.

Phillips, C. W. 1934. The present state of archaeology in Lincolnshire. *Arch. Journ.* **91,** 97–187.

—— (ed.). 1970. *The Fenland in Roman Times*. London.

Richmond. I. A. 1963. *Roman Britain.* London.

Rivet, A. L. F. 1964. *Town and Country in Roman Britain.* 2nd edn. London.

Robson J. D., George, H. and Heaven, F. W. 1974. *Soils in Lincolnshire 1 Sheet TF16 (Woodhall Spa).* Harpenden.

Schadla-Hall, R. T. and Hinchcliffe, J. (eds.). *Plough Damage and Archaeology* (forthcoming in DoE Monograph).

Simmons, B. B. 1975a. The Lincolnshire Fens and Fen edge north of Bourne. Unpublished M. A. dissertation, University of Leicester.

—— 1975b. *The Lincolnshire Car Dyke.* Swineshead.

—— 1975c. The excavation of an Iron Age salt-making site, Helpringham, Lincolnshire (forthcoming).

—— 1975d. A Roman salt-making site and possible sea bank in Aslackby Fen, Lincolnshire (forthcoming).

—— 1979. The Lincolnshire Car Dyke: navigation or drainage? *Britannia,* **10,** 183–96.

—— and Chowne, P. 1975. A section through Horbling Fen, Lincolnshire (forthcoming).

Swinnerton, H. M. 1932. The pottery sites of the Lincolnshire coast. *Antiq. Journ.,* **12,** 239–53.

Todd, M. 1973. *The Coritani*. London.
Tooley, M. J. 1974. Sea-level changes during the last 9000 years in north-west England. *Geog. Journ.*, **140,** 18–42.
Trollope, E. 1872. *Sleaford and the Wapentakes of Flaxwell and Aswardhurn*. London.
Whitwell, J. B. 1970. *Roman Lincolnshire*. Lincoln.

REFERENCES

1 *The Wash Storage Scheme. A Historical Review* (1975, Wallingford). The report is not intended to be used for general publication.
2 Obtained from Soil Survey of England and Wales, Kesteven C.C. (now North and South Kesteven D.C.), Eastern Airviews and R.A.F.
3 Fieldnotes made by D. Robson, Soil Survey of England and Wales.
4 Pottery now in possession of Lincoln Museum.
5 Mainly in Lincoln.
6 Information from Miss R. H. Healey, South Lincolnshire Archaeological Unit.
7 Information from Mrs G. Wilson.
8 A radiocarbon date of the layer immediately beneath the hearth (the layer containing seeds of seablite) would seem to suggest that the influence of the sea was experienced here at about 200 bc.
9 Fieldnotes, record cards and drawings in possession of South Lincolnshire Archaeological Unit.
10 Information from R. Evans, Soil Survey of England and Wales.
11 Information from D. Robson and R. Evans, Soil Survey of England and Wales.
12 Drainage Boards' redrawn plans and sections in possession of South Lincolnshire Archaeological Unit.
13 Carbon dates from Horbling have already been obtained.
14 I am grateful to H. Hurst for sight of his unpublished report on excavations on the Roman defences at Horncastle, and for discussing his observations on the possibility of Horncastle being part of the late Roman defences around the Wash.

# Theories of Coastal Change in North-West England

Dr M. J. Tooley

### *Introduction*

The present coast of north-west England is in part an upland and cliff coast and in part a low-lying coast. Cliffs and uplands abutting the coast are a characteristic of Cumbria south of Maryport, but in Lancashire, south of Morecambe Bay and north of Maryport in Cumbria, much of the coast and its hinterland is low-lying.

The low-lying part of the coast comprises different geomorphological landforms:

1. Extensive coastal flats with sand banks exposed during low water, where sand, silt and clay, either the products of coastal erosion, or delivered from drainage basins or carried landward from the Irish Sea, are moved by tidal currents and waves during the tidal cycle. The ridge and runnel morphology of the Fylde and south-west Lancashire coasts is a well-known characteristic form (Gresswell 1953; King 1972; Parker 1975). The intertidal zone here attains maximum widths of about 4 km (2.48 miles) reflecting both sediment supply, submarine morphology and wide tidal ranges along the Lancashire and Cumbrian coasts.

2. From this broad intertidal zone, sand has been recruited to build the extensive belt of sand dunes along the Southport and Blackpool coasts. The dune belt is about 4 km (2.48 miles) wide at its maximum and dunes rise to *c.* 30 m (98 ft.) above sea-level in the Birkdale Hills. The dunes along the southern shore of the Solway are not as extensive, and are localised further south at Drigg and Eskmeals.

3. Flanking the dunes are salt marshes as the estuaries of the Mersey, Alt, Ribble and Solway are approached. In the Ribble estuary, there are 1730 ha (*c.* 4274 acres) of unreclaimed marsh, and the salt marsh plant communities are characterised by *Puccinellia maritima* (King 1976). The altitude of the salt

marshes varies from + 2.8 to + 4.8 m O.D. (9.1–15.7 ft.) (Tooley 1978a). In Morecambe Bay, mature, grazed salt marshes have an altitudinal range of + 5.3 to + 6.6 m O.D. (17.3–21.6 ft.) (Gray and Bunce 1972).

4. Behind the coastal dunes and salt marshes, many of which have been embanked and reclaimed, are extensive areas of farm land, at present well below the High Water Mark of Spring Tides. In the south-west Fylde there are about 54 sq. km (20.84 sq. miles) below + 7 m O.D. (22.9 ft.) and about 148 sq. km (57.12 sq. miles) in south-west Lancashire.

Within this low-lying region, four distinctive palaeoenvironments are recognised: the perimarine zone, the tidal flat and lagoonal zone, the sand dune zone and the saltmarsh zone (after Hageman 1969). Each zone has a characteristic litho- and biostratigraphy. The perimarine zone is an area in which freshwater clays and peats accumulated: there are no estuarine or marine facies recorded but movements of sea-level to seaward have affected the sedimentary history by affecting the height of the water-table. The tidal flat and lagoonal zone is an area in which marine to brackish water clays, silts and sands alternate in the stratigraphic column with telmatic and terrestrial peats.

The litho- and biostratigraphy in these zones has been interpreted differently by various authors, and this is the basis for a consideration of theories of coastal change in north-west England.

## *Theories of Coastal Change*

Coastal deposits in south-west Lancashire have attracted attention and comment for the past three hundred years, and there has been speculation about the origin of the 'submerged forests', the alternating layers of peat and silt beneath the mosses and the burning dykes on Downholland Moss. Three theories have been advanced to explain these features by Binney and Talbot (1843), Reade (1871) and Gresswell (1953, 1957).

Binney and Talbot (1843) made an excavation on Downholland Moss and identified two organic horizons intercalating blue, sandy, silty clay. They interpreted the minerogenic layers as marine transgressions caused by the breaching of sand barriers and the organic horizons as the result of marine regressions caused by the shoaling of sand in tidal inlets and impeded drainage. No land or sea-level movements were invoked, but a periodic breaching of sand barriers and shoaling in tidal inlets were advanced as an explanation.

Reade (1871) identified four lithostratigraphic units: an Inferior Peat and Forest Bed; a Marine clay (the Formby and Leasowe Marine Beds); a Superior Peat and Forest Bed, and recent silts. He explained the deposition of the Marine clay in terms of subsidence and a consequential marine transgression, whereas the 'Inferior' and 'Superior' Peats were explained as a consequence of uplift. In order to explain the complete sequence of unconsolidated deposits in south-west Lancashire, Reade recognised two periods of uplift and three periods of subsidence.

Gresswell (1953, 1957) reinterpreted the lithostratigraphic successions in south-west Lancashire. He argued cogently for a rise in sea-level to at least + 5.7 m O.D. (18.7 ft.) about 6000 years ago. During this period of high sea-level, an extensive transgression occurred and this part of the dilated Irish Sea was called the 'Hillhouse Sea'. Along the landward margin of this 'sea' a cliff was cut in the boulder clay at + 5.1 m O.D. (16.7 ft.), and to this erosional feature Gresswell gave the name, 'Hillhouse Coastline' after the type site at Hillhouse. After this high sea-level stand, sea-level fell to − 8.2 m O.D. (−26.9 ft.) because of continued land uplift, and, as it fell, on a prograding shore, so sand dunes (Shirdley Hill Sand), then marine clay and silt (Downholland Silt) with intercalated organic deposits formed, and along the coast, at a lowered altitude, sand dunes accumulated. Finally, Gresswell argued for a rise in relative sea-level, eroding the frontal dunes and pushing the dune belt landward over the seaward margin of the mosslands.

In summary, three extant theories had been put forward to explain the distribution of coastal landforms and lithostratigraphy in the tidal flat and lagoonal zone: (1) the breaching of sand barriers and the shoaling of inlets during a period of static sea-level; (2) land subsidence and uplift during a period of static sea-level, but with relative rises and falls of sea-level; (3) an oscillating sea-level caused by the interplay of land uplift and the eustatic rise of sea-level.

These theories can be tested by examining the bio- and lithostratigraphy at classic sites in south-west Lancashire.

## *Evidence of Coastal Palaeoenvironments in South-West Lancashire*

There are many sites in south-west Lancashire from the perimarine zone, the tidal flat and lagoonal zone, and the sand dunes where borings and excavations reveal evidence of coastal palaeoenvironments and sea-level changes both indirectly and directly. A summary of some of the evidence is given in the succeeding sections and more detailed site descriptions can be found elsewhere (Tooley 1974, 1976, 1977, 1978 a,b,c).

The classic site in south-west Lancashire is Downholland Moss (SD 3208). West of Old Moss Lane and along the length of Downholland Moss Lane, the stratigraphy indicates a typical tidal flat and lagoonal palaeoenvironment: mono cotyledonous *turfa* and *limus* deposits intercalate marine and estuarine *argilla* and *grana* (see figs. 6 and 7 in Tooley 1978a). On Downholland Moss, three major marine transgressions are recognised. The first occurred from 6980 to 6755 bp and there are altitudinal limits of −0.72 ± 0.47 to −0.19 ± 0.29 m O.D. (*c.* −2.36 to −0.62 ft.) on the transgressive and regressive contacts of the transgression. The second transgression occurred from 6500 to 6050 bp and there are altitudinal limits of +0.33 ± 0.18 to + 1.07 ± 0.50 m (*c.* 1.08–3.51 ft.). The third transgression occurred from 5900 to 5615 bp and the altitudinal limits are +0.86 ± 0.31 to +1.80 ± 0.38 m (*c.* 2.82–5.90 ft.). Although the altitudinal differences are slight—in part a consequence of post-depositional consolidation — the transgressions are individualised by altitude and by age, indicated relatively by the pollen assemblage zones from the intercalated biogenic sediments and 'absolutely' by the radiocarbon dates. Micropalaeontological and sedimentological analyses indicate fundamental changes in water depth and water quality during the sedimentation both of the biogenic and minerogenic sediments (Tooley 1978).

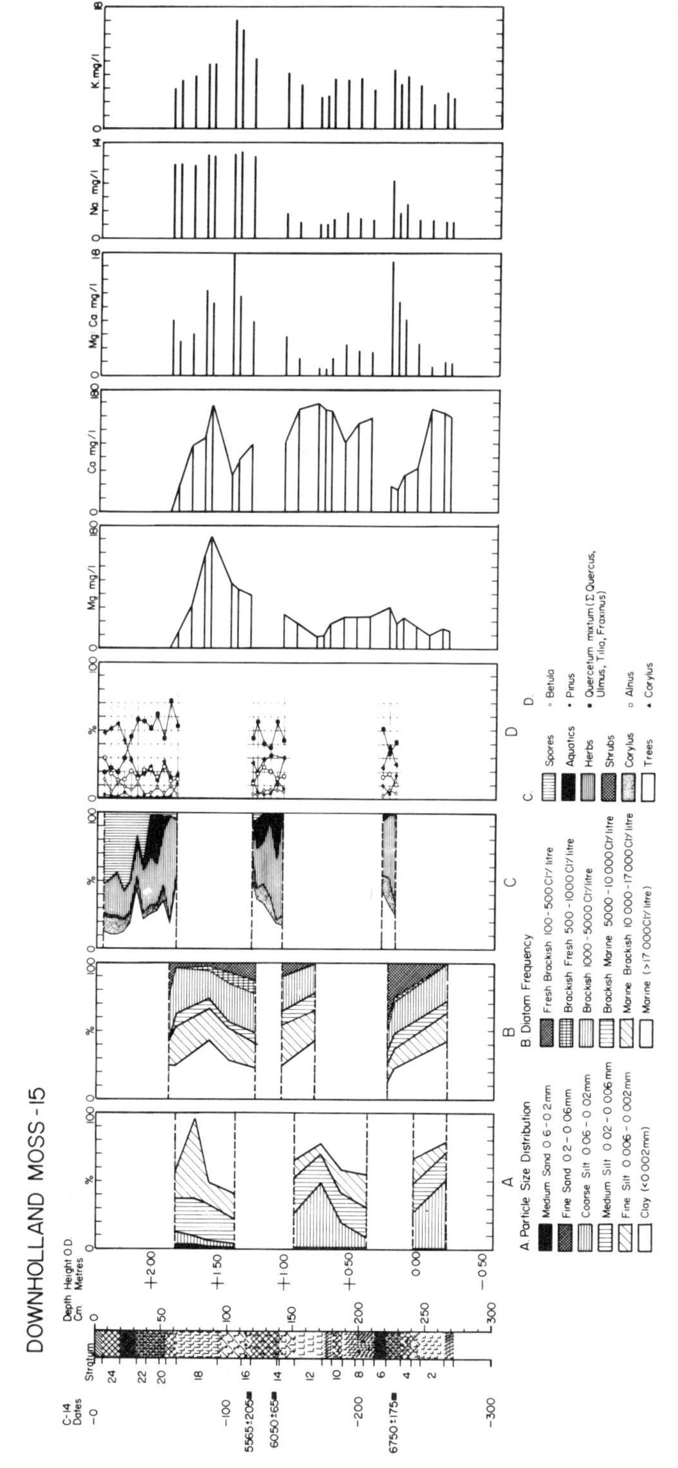


Fig. 34. Downholland Moss–15. Stratigraphic, radiometric, granulometric, diatom, pollen and chemical analyses from undisturbed cores taken with a Proline corer. The stratigraphic symbols are according to Troels-Smith (1955).

For example at Downholland Moss–15 (SD. 3202 0838) three marine episodes are recognised. Episode 1 occurred whilst Strata 1–3 (fig. 34) from 271 to 241 cm (106.6 to 94.8 in.) were accumulating: the succession from a silt with sand laminations, to a silt with gyttja, sand laminations and iron-stained root channels points to a change from an upper tidal flat to a salt-marsh palaeoenvironment. The decline in the frequency of marine diatoms and of the particle size as the 3/4 boundary is approached supports this interpretation. A relative fall of sea-level is inferred. Episode 2 is recorded by Strata 7–13 from 205 to 140 cm (80.7–55.1 in.). Stratum 7, which comprises a silt with iron staining and discrete gyttja partings, is interpreted as a salt marsh palaeosol. This gives way to a silt and silty clay, with stress structures and no laminations, which indicates a transition to higher mud flats. The laminated sandy silts and silty clays of Strata 11 and 12 indicate the lower part of the higher mud flats. The progressive change in sedimentary environment is a consequence of a rising sea-level. The succeeding strata (12–14) indicate a fall in sea-level as laminated sandy silts are replaced by clay gyttjas with iron staining in root channels, pointing unequivocally to a change in sedimentary environment from higher mud flats to salt marshes. A similar succession is recorded during Episode 3. Foraminiferal analyses from sediments referable to these three episodes confirm these interpretations (Huddart, unpublished).

The intercalated biogenic sediments contain stratigraphic and microfossil evidence of an autogenic plant succession followed by a retrogressive one: from saltmarsh communities to reed swamps, freshwater aquatic communities and an oak-dominated fen, back through aquatic communities and reed swamps to salt marshes.

In the perimarine zone of Downholland Moss east of Old Moss Lane a fluctuating watertable and a consequential change of sediment type from terrestrial to telmatic and limnic biogenic deposits and back to terrestrial deposits are closely related to marine transgressions recorded further west in the tidal flat and lagoonal zones. For example, at Downholland Moss–11 (Tooley 1969), the first peak in the frequency of the pollen of bullrushes (*Typha angustifolia*) occurs at an altitude of *c.* −0.18 m O.D. (−0.59 ft.) which falls within the altitudinal limits of the first transgression on Downholland Moss (see p. 76). The second peak of *Typha* pollen occurs from +0.22 to +0.32 m O.D. (0.72–1.04 ft.) which is within the limits set for the second transgression. The third transgression affected sites only 300 m (984 ft.) west of DM–11 where there was a recrudesence of freshwater plant communities, and, in addition, a reappearance of the pollen of salt marsh taxa, such as Chenopodiaceae, *Artemisia* and *Armeria* at altitudes of +0.42 to +0.72 m O.D. (1.37–2.36 ft.). This is lower than the altitudes of the third transgression, and it is probable that altitudes at DM–11, that comprised in 1968 313 cm of biogenic sediments above a wedge of marine clay 51 cm (20.0 in.) thick, have been affected by post-depositional consolidation.

In the present-day intertidal zone off the south-west Lancashire and Fylde coasts peat beds have been recorded from time to time. Some of these so-called 'submerged forests' are related to former positions of sea-level, whereas others are not, and only micropalaeontological analyses can indicate proximity to the sea or not (see discussion in Tooley 1978a, 176–7). The peat beds at Rossall and Cleveleys are basal organic deposits from kettle holes, the circumvallating ramparts of which have been removed by coastal erosion probably during Flandrian

III, that is during the last 5000 years. The peat beds at the Altmouth (Tooley 1970, 1977) and south of Formby Point are related to former sea-level stands.

Seaward of the end of Lifeboat Road and south of Formby Point, a sandy, woody detrital peat has been recorded from the upper intertidal zone at an altitude of +5.18 to +5.08 m O.D. (16.99–16.66 ft.) (Tooley, unpublished). A radiocarbon assay yielded a date of 2335 ± 120 (Hv. 4709). The peat was underlain by bleached white sand, and until recently (pre-1968) was overlain by dune sand. The organic deposit is interpreted as a fossil dune slack, and, on the basis that biogenic sedimentation occurs in slacks during marine transgression episodes (Jelgersma *et al.* 1970), the deposit at Formby can be interpreted thus. In the Fylde, a marine transgression was under way from 3090–2270 bp (Tooley 1978a) and this lends support to the interpretation of Jelgersma *et al*. Lower on the beach, an area of silt poached by cattle has been recorded. It is probable that the hoofprints are those of domesticated oxen (J. Jewell 1976, personal communication) and, being at a lower altitude, beneath the dated level, can be tentatively referred to the Iron Age.

Further evidence of the age of the dunes and the relationship to former sea-level stands comes from the landward limit of the Formby Hills in south-west Lancashire and the Starr Hills in the south-west Fylde. At the former, sand has overblown the seaward margin of Downholland Moss at an altitude of *c.* +3.9 m O.D. (12.79 ft.) and at DM–3c a radiocarbon assay on monocotyledonous *turfa* subjacent to the sand yielded at a date of 4090 ± 170 (Hv. 4705). The interpretation of these data is that shortly after 4090 bp sand dunes began to accumulate in this area, and according to the model of Jelgersma *et al.* (1970) this period of dune building and unstable sand should correspond with a period of marine regression. In north-west England, a period of relatively low sea-level is recorded between Lytham Vla and Lytham VIII from 4545 to 3700 bp (Huddart *et al.* 1977). In the Netherlands, the oldest period of dune building was completed by 4100 bp towards the end of Calais IV transgression (Jelgersma *et al.* 1970; Oele 1977), and this suggests a synchronism of dune-building episodes.

A period of dune stability in the Lancashire dunes is indicated by the date of 2335 bp from Formby Foreshore and this is supported by a date on juniper wood embedded in sand from Velsen–Hoogovens of 2450±40 separating two dune-building episodes in the Netherlands (Jelgersma *et al.* 1970).

A final dune-building stability cycle is recorded from the Dark Ages. The Starr Hills have a well-developed biogenic horizon, which in Lytham Hall Park attains a thickness of 12 cm (4.72 in.). This layer, interpreted as a fossil dune slack deposit (see Tooley 1978a), has been dated to bp 805 ± 70 (Hv. 4417) and bp 830 ± 50 (Hv. 3846). If this period of dune stability is associated with a marine transgression then the sandy facies below and above point to periods of marine regression. That the dunes were unstable during these periods is confirmed by documentary evidence from the chartulary of Lytham Priory. Furthermore, the dune chronology established in the Netherlands indicates a period of dune building shortly after 800 bp and assigned to the twelfth and thirteenth centuries.

The evidence from the perimarine, tidal flat and lagoonal and sand dune zones of south-west Lancashire points to a repetitive pattern of marine transgressions: each transgression, recorded in the tidal flat and lagoonal zone, pushed landward to a different extent and caused either a dilation or contraction in the extent of the perimarine zone. A further consequence was a change in the ordination of the plant communities in the coastal area and a change in the extent and location of biotic resources. The implications of a series of marine transgression sequences in terms of sea-level movements is considered in the succeeding section.

### *Sea-Level Movements*

A marine transgression, recorded directly in the tidal flat and lagoonal zone, comprises a transgressive contact in which minerogenic sediments overlie biogenic sediments and a regressive contact in which biogenic sediments overlie minerogenic sediments. The minerogenic sediments comprise sands, silts or clays, sometimes laminated, occasionally bioturbated and enriched with ferric oxide in root channels or as a diffuse iron stain. For each transgression recognised in north-west England (Lytham I to IX), a characteristic facies has been described (see Table 1, pp. 136–7, in Huddart *et al*. 1977). The boundaries between the transgressive and regressive contact and the underlying or overlying biogenic material tend to be sharp and eroded, with flames of biogenic material pushing into the minerogenic material for the former and transitional and uneroded for the latter.

A rise and fall of sea-level is comprehended by a single marine transgression. The transgressive contact records the progressive rise of sea-level and facies changes within the minerogenic layer, as described earlier (p. 78) and elsewhere (Tooley 1978b), record a progressive increase in water depth in relation to the tidal cycle as the tidal range moves with the rising sea-level. Conversely, the regressive contact records the progressive fall of sea level, and the removal of the marine effect is presaged by unequivocal indicators of reducing water depth and period of tidal immersion as the regressive contact is approached.

The consistent pattern of sedimentation during a marine transgression episode, indicated by micropalaeontological, sedimentological and chemical evidence, militates against an explanation that invokes breaching of natural defences during periods of storminess. If this mechanism operated, then sheets of coarse sediment and erosional surfaces could be expected, yet none is recorded. The 1953 storm surge in eastern England left little sedimentary evidence: at Scolt Head there was limited deposition of clay and sand after the breach made in 1953 (J.A. Steers 1976, personal communication). But the sedimentary record of storm surges cannot be invoked to explain the extensive marine transgressions recorded in the tidal flat and lagoonal zone (the first transgression on Downholland Moss extended landward some 7 km (4.34 miles) from the present coast), in which progressive and retrogressive changes in water depth and water quality have been demonstrated and radiometric time limits, corroborated pollen-analytically, indicate periods for these marine events of 170 to 800 years.

Furthermore, Professor R. W. Fairbridge showed most eloquently that in middle latitudes an increase in storminess would be associated with marine regressions consequent upon a build up of high latitude ice caps and shelf ice, and more southerly tracks for low pressure cells. A fall of sea-level would extend the intertidal zone, and wave energy would be considerably reduced during a period of storminess, having to cross this dilated zone. Breaching of natural sea-defences during these periods would thereby be reduced.

It is possible, therefore, that the transgressive part of a marine transgression comprehends a rise of sea-level and that the regressive part of a transgression comprehends a fall of sea-level, as explained in 1974 (Tooley 1974, 33). Support is lent to this interpretation by the near-synchronism of marine events in both uplifted and subsiding areas, that demonstrates the over-riding control of the sea-level oscillation. For example, in the Forth valley of Scotland, an uplifted area, the onset of marine conditions and the beginning of the deposition of Carse clay is dated (bp) to 8270 ± 140 (I. 1838), on the Fylde coast of Lancashire at 8390 ± 105 (Hv. 4343) and in Morecambe Bay at 8330 ± 125 (Hv. 3462). An earlier fall of sea-level is dated (bp) in the Forth to 8690 ± 140 (I. 1839) and in the Fylde to 8575 ± 105 (Hv. 4346).

In the Netherlands and particularly in south-east England, there is a marked marine regression towards the end of Flandrian II and the beginning of Flandrian III (the Atlantic/Sub-Boreal transition), and this regression is a feature of north-west England. In north-west England dates (bp) of 5250 ± 385, 5015 ± 100,4960 ± 50,5435 ± 105,5005 ± 65,4800 ± 75,4830 ± 140,4845 ± 100, i.a., record the end of marine conditions (see Table 1, in Tooley 1978a). In the Lower Thames estuary Devoy (1977) has given similar dates (bp) for the end of marine condition: on Broadness Marsh 5220 ± 65 (Q. 1341), at West Thurrock 4975 ± 120 (IGS/152) and on Stone Marsh 4930 ± 110 (Q. 1136). In the Netherlands Jelgersma (1961) has given dates (bp) for the end of marine transgressions (corrected for the Suess effect, according to the table given in Vogel and Waterbolk 1963) such as 4920 ± 65 (GrN−1113: Waarde II), 5420 ± 70 (GrN−1143: Zuidland) and 5200 ± 140 (GrN−222: Willemstad II).

An earlier Flandrian II marine regression is recorded at Nancy's Bay in Lancashire at 6290 ± 85, and by Devoy (1977) at Tilbury (World's End) at 6200 ± 90 (Q. 1430).

Near Montrose in north-east Scotland, Smith *et al*. (1978) have dated the beginning of marine sedimentation at Fullerton to 7140 ± 50 and 7086 ± 50, which is close to the onset of Lytham IV in north-west England, where dates of 6980 ± 55 and 6885 ± 80 record the time of inception of the transgression in south-west Lancashire and the Fylde.

If sea-level was rising smoothly (see Kidson and Heyworth 1973; Jelgersma 1961, 1966), it is unlikely that in subsiding areas such as the Netherlands and south-east England, marine regressions occurring here and in north-west England at about the same time would be recorded other than as a function of an absolute fall in the sea-level surface. By the same token, it is unlikely that in uplift areas, such as Scotland, marine transgressions occurring at about the same time would be recorded unless the absolute rise in the sea-level surface was sufficient to transgress the land and leave a marine record.

The inescapable conclusion that the restoration of sea-level during the Flandrian stage was achieved by a series of oscillations and consequential marine transgressions, comprising transgressive and regressive phases.

In north-west England, twenty-six index points have been used to control the period and amplitude of the sea-level oscillations (fig. 35). These index points and the period and amplitude they indicate are corroborated by a further thirty-three points related to former sea-level stands from sites outside the type area of Lytham.

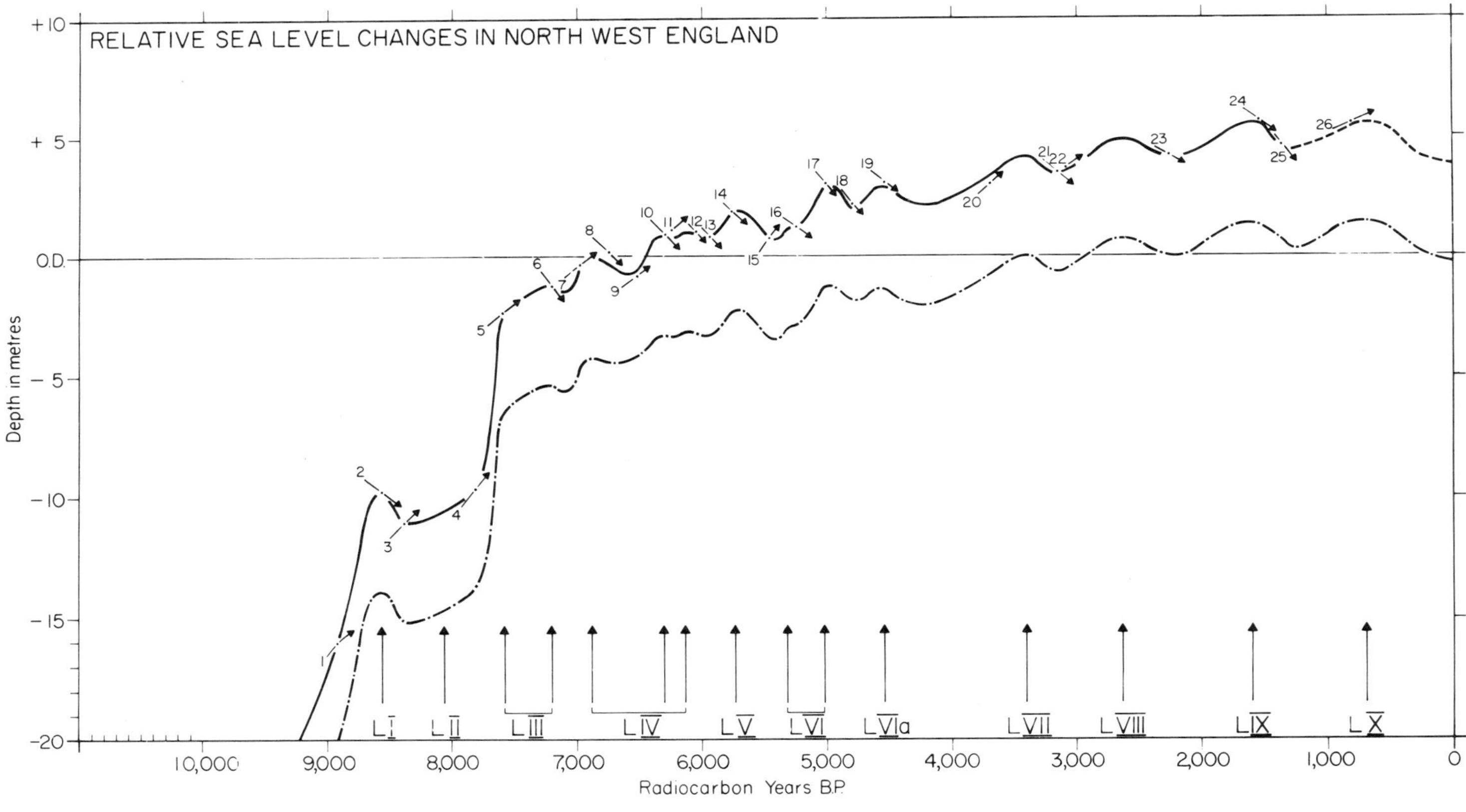


Fig. 35. Relative sea-level changes from a restricted area in west Lancashire based upon twenty-six index points (see Tooley 1978a). The continuous line curve shows the change in altitude of Mean High Water Mark of Spring Tides: the dot-dash curve shows Mean Tide Level, assuming a constant relationship between MHWST and MTL during the Flandrian stage. Vertical arrows LI to LX indicate marine transgression maxima.

## *Sea-Level Changes During the Prehistoric Period in North-West England*

The evidence of sea-level changes during the Palaeolithic in north-west England is fragmentary and speculative. There is some evidence for a Late Devensian marine transgression on the Isle of Man younger than 25000 bp but older than 18900 bp: at the Dog Mills on the east coast of Man (Thomas 1977) clay, silts and laminated sands with a foraminiferal assemblage indicative of cold water estuarine conditions was recorded between lodgement and thin flow tills, at an altitude of about +14 m O.D. (45.93 ft.). At Blackpool, a foraminiferal deposit underlies till and on the Wirral at +18 m O.D. (59.05 ft.), a shelly, marine sand also underlies till. If there was a marine accompaniment to deglaciation in the Irish Sea, as Thomas (1977) proposes, it is unlikely that there would have been habitats sufficiently genial for Palaeolithic folk.

At Kirkhead Cavern in Cartmel, Mellars (1970) has described nine flint artifacts which he assigns to the late upper Palaeolithic, probably about 14000 bp which is consistent with evidence for an ice free Lake District at least by 14300 bp. Although water-eroded notches, interpreted as marine by Ashmead (1974), are recorded at the altitude of Kirkhead Cavern (+30 m O.D. (98.42 ft.)) and at altitudes of +15 (49.21 ft.) and +12 m (39.36 ft.) O.D. on Kirkhead Hill, Humphrey Head and Warton Crag, there is no way at present of establishing the age of these erosional features, and relating them to the period of prehistoric occupation in Kirkhead Cavern.

Evidence of sea-level changes during the Mesolithic period is stronger from 10,000 to 5,000 bp; there were marked changes in the sea-level surface and extensive transgressions both of the continental shelf and landward of the present coast occurred. Evidence from Morecambe Bay indicates that by 9270 ± 200 bp (Birm. 141) a proto-bay had formed, and this must be a minimum date also for the formation of the Irish Sea (Tooley 1978c). Whereas land connections between Ireland, the Isle of Man and mainland Britain appear to have been severed by the beginning of the Flandrian Stage, land connections between Britain and mainland Europe remained open until about 7800 bp (Kolp 1976).

The Fylde coast of Lancashire was transgressed about 8390 bp, and rather significantly marine facies of Lytham II age overlie compact, well-humified woody detrital peat with a dense layer of charcoal some 16 m (*c.* 52 ft.) below the ground surface. The pollen diagram from this site (the Starr Hills, see p. 80, in Tooley 1978) shows peak frequencies of *Corylus, Calluna, Gramineae,* and *Pteridium* at the charcoal horizon and a single grain of *Plantago lanceolata*. Such an assemblage could be interpreted as a product of fire associated with Mesolithic folk.

Occupatiin of these low-lying coastlands was probably short-lived, because the rise of sea-level after 8390 bp was rapid. For example, Lytham III from 7605–7200 bp involved a rise of sea-level of 7 m (22.96 ft.) from −9.6 to −2.3 m O.D. (−31.4 to −7.5 ft.) (Tooley 1978a).

After 7000 bp the rate of overall sea-level rise slackened considerably, but was marked thereafter by a series of oscillations (see fig. 35). The palaeogeography of the coastal hinterland of Lancashire was one of extensive marine transgressions manifested by sheets of sands, silts and clays, separated by limnic, telmatic and terrestrial peats during marine regression stages. The changes in water quality and water depth during the latter half of the Neolithic period would have discouraged occupation of the tidal flat, lagoonal and perimarine zones, and it is probable that Mesolithic folk would have kept to the drier, more open conditions that prevailed east of the +7 m (22.96 ft.) contour where the late-glacial coversand formation —the Shirdley Hill Sands—outcropped.

The opening of the Neolithic period was marked by falling sea-levels and extensive marine regressions. Extensive areas of the former tidal flat zone were exposed to sub-aerial weathering and colonisation by terrestrial plant communities. Clearance of coastal forests of oak is clearly indicated in the pollen diagrams shortly after the end of minerogenic sedimentation.

In north-west England the Bronze Age and Iron Age are characterised by three marine transgressions of limited geographical extent. Transgressions Lytham VIIa, VII and VIII were confined to small embayments such as Lytham Hall Park, the Altmouth, Formby and Crossens. In the Fenlands and in the Netherlands, this period is characterised by extensive marine transgressions. For example, the deposition of Fen Clay is dated from 4690 to 4000 bp, and Calais IV transgression in the Netherlands is dated from 4600 to 3800 bp (Oele 1977).

During the Romano-British period in north-west England, there is evidence for the end of one transgression (Lytham VIII) at 2270 ± 65, and the beginning and end of Lytham IX from 1975 ± 200 until 1370 ± 85 bp. The interesting feature of this transgression is that marine sedimentation is recorded to an altitude of +5.4 m O.D. (17.5 ft.). Hence, Lytham IX is the highest marine record on the Lancashire coast throughout the Flandrian stage. This evidence supports Godwin's claim for a transgression of Romano-British age in the Fenlands after 1875 ± 110 (Q. 549, Godwin and Willis 1961) and in the Severn estuary at some time after 2660 ± 110 (Q. 691, Godwin and Willis 1964). That a marine transgression was under way in the Somerset levels at this time is supported by Hibbert (this volume, pp. 103–5) who has recorded a flooding horizon dated from 1850 to 1450 bp which is rather close to the age of Lytham IX. In the Netherlands, the Dunkirk II transgression has almost identical time limits to Lytham IX, namely 1700 to 1350 bp (Oele 1977).

The occupation of the Roman ports at Skippool on the river Wyre and Ribchester on the river Ribble needs to be considered in light of these data: the Mean High Water Mark at Lytham was *c.* +5.4 m (17.5 ft.) some 1700 years bp which is a metre higher than the present altitude of MHWST.

### *Conclusions*

The evidence presented here and elsewhere (Tooley 1974, 1976, 1977 and 1978a,b) points to a markedly oscillating course for the restoration of sea-level during the Flandrian stage. It has been demonstrated with reference to Downholland Moss that actual movements of the sea-level surface, upwards and downwards, were involved, to which the results of both biological and sedimentological investigations bear witness. Furthermore, oscillations were associated with discrete marine transgressions, each one of which is in its correct stratigraphic position: the oldest marine transgressions are overlaid in turn by progressively younger transgressions. All the transgressions are below the altitude of MHWST except Lytham IX, the regressive contact of which lies unequivocally about a metre above present day MHWST.

There is no evidence that, whatever residual isostatic recovery remained at the opening of the Flandrian stage, the magnitude of uplift exceeded the restoration of sea-level, for each oscillation is nicely recorded at progressively higher altitudes.

The extent of marine transgressions varied: in general Lytham I to VI affected extensive areas whereas Lytham VII to IX were confined to inlets. In south-west Lancashire, the maximum landward limit of marine transgressions is close to the limit of grey clays mapped by de Rance (1869), but the limit never reaches the position or altitude of the 'Hillhouse Coastline' (see Tooley 1976).

It is probably that the theories of coastal evolution in north-west England as proposed by Binney and Talbot (1843), Reade (1871) and Gresswell (1953, 1957) are not correct, and that the stratigraphic record and palaeogeography of the tidal flat and lagoonal zone is explained more realistically by an oscillating sea-level during the Flandrian stage.

## BIBLIOGRAPHY


Ashmead, P. 1974. The caves and karst of the Morecambe Bay area. In Waltham A. C. (ed.), *The Limestone and Caves of North-West England*, 201–26. Newton Abbot, David and Charles.

Binney, E. W. and Talbot. J. H. 1843. *On the petroleum found in the Downholland Moss, near Ormskirk*. Paper read at Fifth Annual General Meeting of the Manchester Geological Society, 6 October 1843.

Devoy, R. J. N. 1977. Flandrian sea-level changes and vegetational history of the Lower Thames estuary. Unpublished Ph.D. dissertation, University of Cambridge.

Godwin, H. and Willis, E. H. 1961. Cambridge University natural radiocarbon measurements III. *Radiocarbon*, **3,** 60–76.

—— and —— 1964. Cambridge University natural radiocarbon measurements VI. *Radiocarbon*, **6,** 116–36.

Gray, A. J. and Bunce, R. G. H. 1972. The ecology of Morecambe Bay. VI. Soils and vegetation of the salt marshes: a multivariate approach. *J. Appl. Ecol.*, **9,** 221–34.

Gresswell, R. K. 1953. *Sandy Shores in South Lancashire: the Geomorphology of South-West Lancashire*. Liverpool University Press.

—— 1957. Hillhouse coastal deposits in south Lancashire. *Manchr. Geol. Journ.* **2,** 60–78.

Hageman, B. P. 1969. Development of the western part of the Netherlands during the Holocene. *Geologie Mijnb.*, **48** (4), 373–88.

Huddart, D. 1977. Foraminifera from Downholland Moss. Unpublished MS.

——, Tooley, M. J. and Carter P. A. 1977. The coasts of north-west England. In Kidson, C. and Tooley, M. J. (eds.), *The Quaternary History of the Irish Sea*. Liverpool, Seel House Press.

Jelgersma, S. 1961. Holocene sea level changes in the Netherlands. *Meded. Geol. Sticht.* C VI, **7,** 1–100.

—— 1966. Sea-level changes during the last 10,000 years. In Sawyer, J. S. (ed.), *World Climate 8000 to 0* B.C., 54–69. Proc. of the International Symposium held at Imperial College, London, 18–19 April 1966. London, Royal Meteorological Society.

——, de Jong, J., Zagwijn, W. H. and van Regteren Altena, J. F. 1970. The coastal dunes of the western Netherlands; geology, vegetational history and archaeology. *Meded. Rijks. geol. Dienst.* N.S., **21,** 93–167.

Kidson, C. and Heyworth, A. 1973. The Flandrian sea-level rise in the Bristol Channel. *Proc. Ussher Soc.*, **2** (6), 565–84.

King, C. A. M. 1972. *Beaches and Coasts*. London, Edward Arnold.

—— 1976. *The Geomorphology of the British Isles: Northern England*. London, Methuen.

Kolp, O. 1976. Submarine Uferterassen der südlichen Ost- und Nordsee als Marken des holozänen Meeresanstiegs und der Überflutungsphasen der Ostsee. *Peterm. Geogr. Mitt.*, **120,** 1–23.

Mellars, P. A. 1970. Flints from Kirkhead Cave. *Archaeological News Bulletin for Northumberland, Cumberland and Westmorland*, **8,** 6.

Oele, O. 1977. The Holocene of the western Netherlands. In Paepe, R. (ed.), *Southern Shores of the North Sea*, 50–6. Guidebook for Excursion C17. X INQUA Congress. Norwich, Geoabstracts.

Parker, W. R. 1975. Sediment mobility and erosion on a multibarred foreshore (south-west Lancashire, U.K.). In Hails, J. and Carr, A. (eds.), *Nearshore Sediment Dynamics and Sedimentation*, 151–78. London.

Rance, C. E. de 1869. The geology of the country between Liverpool and Southport. *Geol. Surv. U.K.* London, H.M.S.O.

Reade, T. M. 1871. The geology and physics of the post-glacial period, as shown in deposits and organic remains in Lancashire and Cheshire. *Proc. L'pool. Geol. Soc.*, **2,** 36–88.

Smith, D. E., Morrison, J., Cullingford, R. A. and Jones, R. L. 1978. The age of the main post-glacial shoreline in an area near Montrose, Scotland. Abstract. *Timescales in Geomorphology*. British Geomorphological Research Group Symposium, Hull.

Thomas, G. S. P. 1977. The Quaternary of the Isle of Man. In Kidson, C. and Tooley, M. J. (eds.), *The Quaternary History of the Irish Sea*, 155–78. Liverpool, Seel House Press.

Tooley, M. J. 1969. Sea-level changes and the development of coastal plant communities during the Flandrian in Lancashire and adjacent areas. Unpublished Ph.D. thesis, University of Lancaster.

—— 1970. The peat beds of the south-west Lancashire coast. *Nature in Lancashire*, **1,** 19–26.

—— 1974. Sea-level changes during the last 9000 years in north-west England. *Geog. Journ.*, **140** (1), 18–42.

—— 1976. Flandrian sea-level changes in west Lancashire and their implications for the 'Hillhouse coastline'. *Geol. Journ.*, **11** (2), 37–52.

—— (ed.) 1977. *The Isle of Man, Lancashire Coast and Lake District*. Guidebook for Excursion A4. X INQUA Congress. Norwich, Geoabstracts.

—— 1978a. *Sea-level Changes: North-West England during the Flandrian Stage*. Oxford, Clarendon Press.

—— 1978b. Holocene sea-level changes: problems of interpretation. *Geol. för. Stockh. Förh.*

—— 1978c. Flandrian sea-level changes and vegetational history of the Isle of Man: a review. In Davey, P. (ed.) *Man and Environment in the Isle of Man*, 15–24. British Archaeological Reports (British Series), **54.**

Troels-Smith, J. 1955. Karakterisering af løse jordarter. *Danm. geol. Unders*. IV- Raekke, Bd. 3.

Vogel, J. C. and Waterbolk, H. T. 1963. Groningen radiocarbon dates IV. *Radiocarbon*, **5,** 163–202.

# Archaeology and Coastal Change in the North-West[1]

Professor G. D. B. Jones, F.S.A.

The North-West is an area that archaeologically has received little synthesised study on a broad canvas. Yet in terms of coastal change and its effect on settlement location the Solway–Liverpool Bay area is potentially one of the most interesting in the country, notably since important recent work on sea-level change has been conducted in this area. The following chapter is simply a statement of where archaeological research may profitably be directed. The studies of palaeobotanists in recent years have helped clarify the sequential development of vegetation that recolonised Britain after the Devensian glaciation. At that time the north received a succession of botanical colonists from refuge areas and during the Sub-Boreal period, Zone VI of the palaeobotanists, drier conditions promoted the growth of pine at the expense of birch. In the subsequent wetter Atlantic period from approximately 7500 bp (5550 B.C.) in the lower northern zones oak and elm formed a climax forest, while on high ground birch became dominant in a cover of birch and pine.

Yet the discovery of tree-stumps under subsequent blanket peat, particularly in Scotland, has long attested substantial change, either of climate and/or soil fertility.[2] This raises the question of when the development of blanket peat began and how it was related to climatic change that would in turn have related to absolute sea-level change. In parts of Ireland, on the Pennines, in Aberdeenshire and in the Lake District blanket peat began growing 2200 and 3550 bp.[3] The particular point of this is that in certain places in the North the pollen of other species in the area has been silted and preserved for study. The northern lowland vegetational succession to oak forest is known at Aros Moss on Kintyre, while at Racks Moss, part of Lochar Moss near Dumfries, after an initial rise in alder pollen, indicating the onset of wetter conditions, a layer of estuarine silting shows a temporary imbalance between the eustatic rise of sea-level due to water returning into circulation from the ice caps and glaciers and the isostatic rise of land formerly depressed by ice. These results were published by Nichols in 1968 and have been followed by further work from Jardine and Morrison on the Holocene coastal

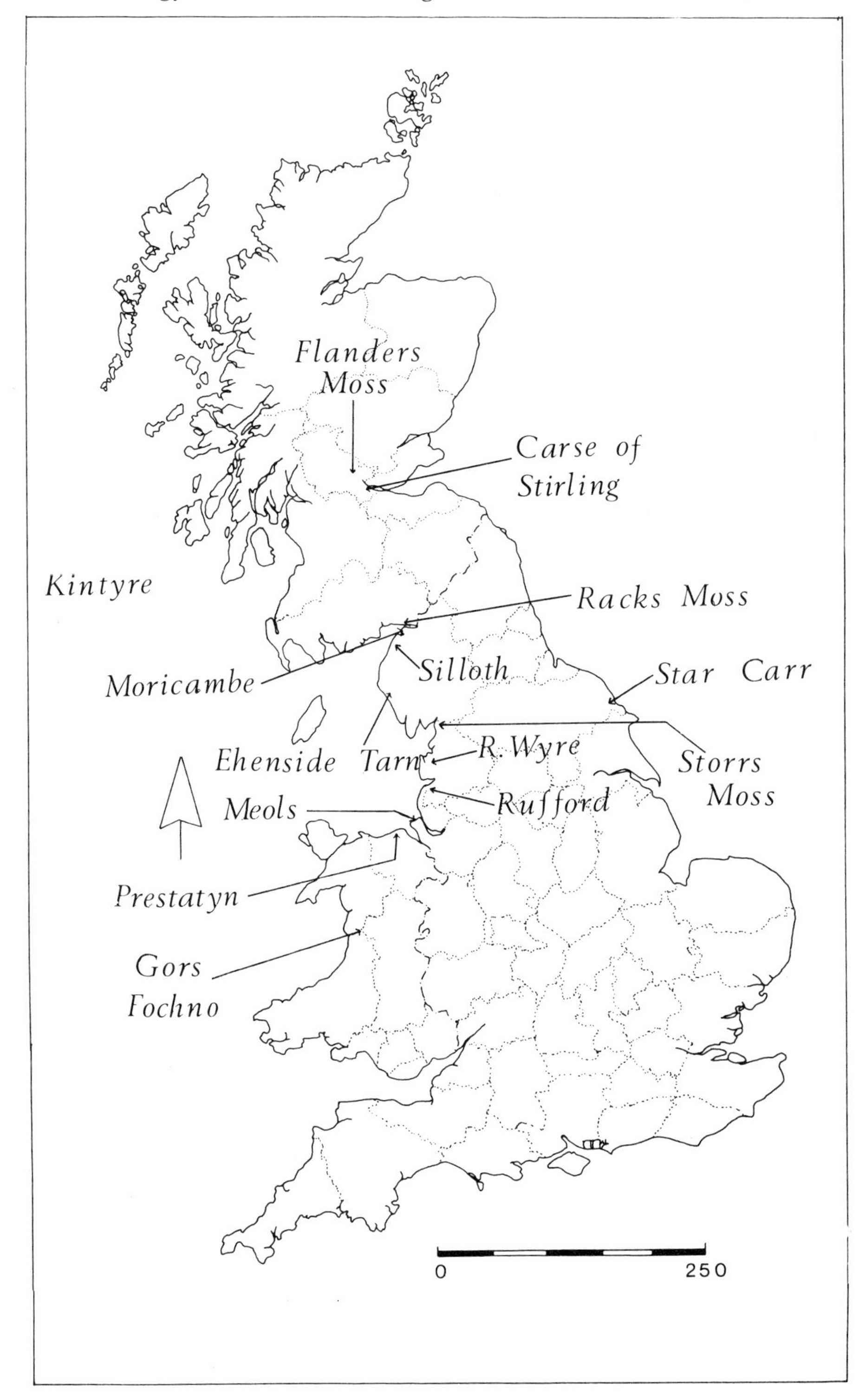


Fig. 36. Sites mentioned in the text.

deposits of south-western Scotland.[4] In particular they have elaborated our knowledge of estuarine sediments in Wigtown Bay and have established the major differences between the present coastal configuration of south-western Scotland and the varying extent of the Holocene Sea at approximately 7200, 6600 and 5600 bp. In this context the establishment of the presence of estuarine deposits in what is now Lochars Moss and the associated work at Racks Moss is of particular importance. In 1973 a dugout canoe was discovered on the edge of Racks Moss and dated by radiocarbon to 3754 ± 125 bp, a period which suggests the survival of lagoon conditions into the late Neolithic/Early Bronze Age.

The temporary submergence of part of the coast of the Solway Firth may have been more or less contemporary with post-glacial marine transgression across part of the Carse of Stirling. Both areas became freshwater lakes as the land continued to rise and were subsequently converted into fen carr or sphagnum moss during the Sub-Boreal period. The work of Jardine and Morrison has isolated a number of mesolithic settlements in Galloway, but for later periods the evidence is scant, though an air photographic programme currently in progress may help rectify the imbalance. Across the south Solway marsh settlements, perhaps of a more permanent nature, have been found at Ehenside Tarn on the Cumberland coast and Storrs Moss in Morecambe Bay, where neolithic forest clearance began early, about 5400 bp, on the coastal lowlands.[5]

The work of Jardine and Morrison shows that by the late Iron Age period the north Solway shore had achieved roughly its present configuration, reflected in the latest air photographic distribution maps for Iron Age sites along the edge of Lochars Moss. By a fortunate coincidence, when we turn to the south side of the Solway, the recent progressive discovery of linear defences, forming a western extension of the Hadrian's Wall system, provides a reasonably precise indicator of the coastline. In particular, the ditches and palisades now known to run along the coastal strip west of Bowness and across Moricambe enable the shape of the coastal headlands to be reconstructed in some detail for the second century A.D.[6]

The discovery in 1975 of parallel ditches linking the milefortlets known to stretch along the Solway Estuary has considerable ramifications for our understanding of the immediate layout of the present coastline. This is particularly the case against a background in which it has usually been assumed that large-scale flooding in the thirteenth century altered the coastal alignment in the Skinburness area, the Edwardian naval base on the south side of Moricambe, the joint estuary of the Waver and Wampool.[7] In fact the ditched cordon linking the milefortlets and now known to contain, first, timber towers with an associated palisade and, later, stone towers suggests that there was relatively little change in the actual shoreline. Normally the cordon can be shown to have run along the first or second of the raised beaches between Bowness and Cardurnock. The alignment of the cordon suggests marginal coastal alignment west of Milefortlet 1 (Biglands); it also shows the way in which the foreland has been eroded at North Plain in a way that makes it likely the Milefortlet 2 has been completely lost. The detailed evidence from the sections cut across the ditches close to Tower 2B shows the occurrence of flood incursion sometime during or after the Roman period. The exact position is given in fig. 37 and shows that the greatest storm-driven high tide reached a height some 4.8 m (16 ft.) above the present high water levels at spring tides. Naturally the easiest explanation is that some of this additional height is

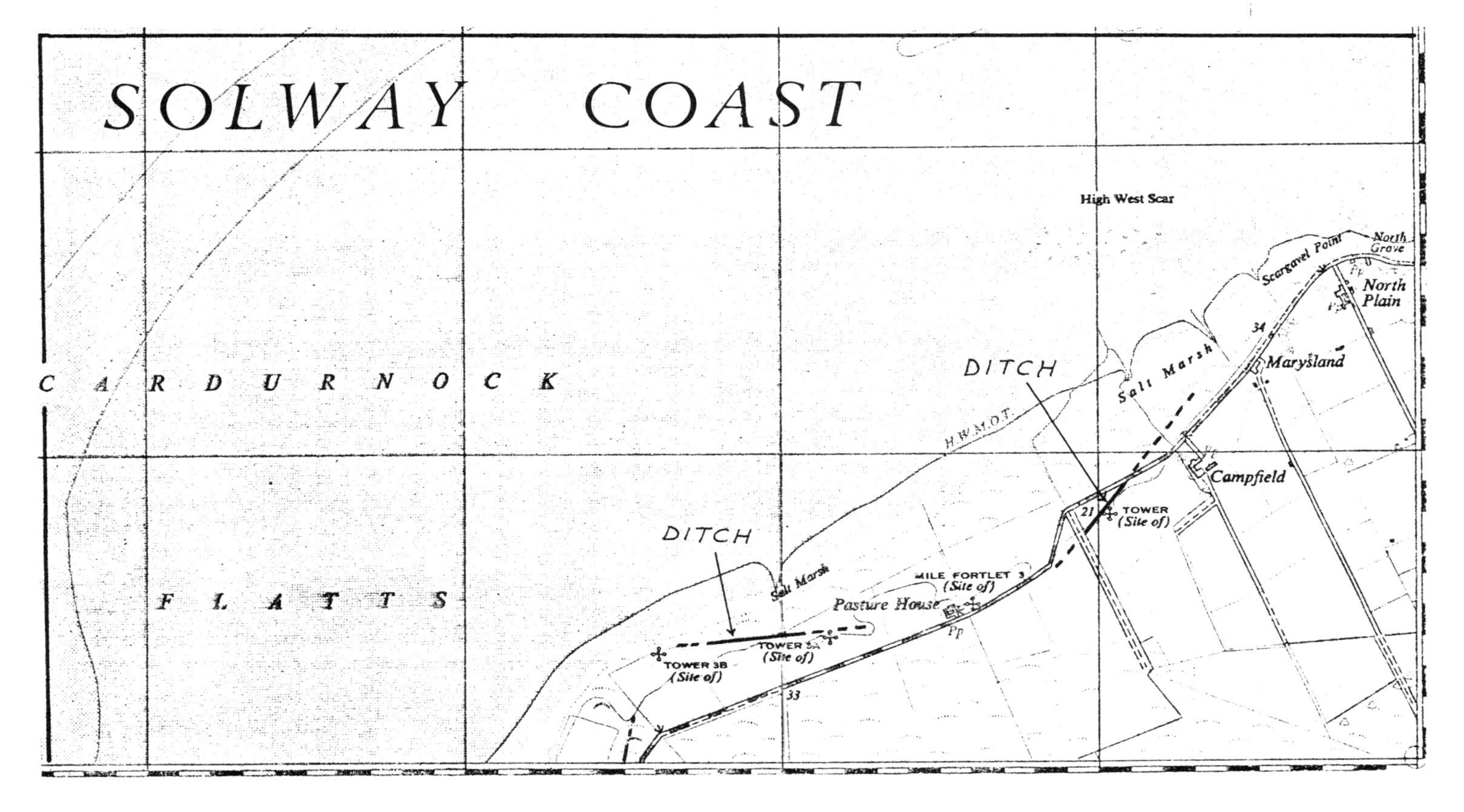


Fig. 37. Solway coast.

directly related to the postulated rise in tidal ranges towards the end of the Roman period and afterwards. On the south side of Moricambe the continuation of the palisade system suggests relatively little change in the southern area. Further south again, at Dubmill Point, the recent discovery of Milefortlet 17 shows, as had not been expected, that the milefortlet survives on the present headland and had suffered only mild erosion. A roughly similar situation is known at Beckfort where the well-known fort directly over-looking the present foreshore replaced an earlier milefortlet that now lies partly eroded amongst the present dunes. A similar replacement of milefortlet by fort is visible at Maryport where the former has been eroded in the present cliff face, but the position of neither has a direct bearing on the problems of absolute sea level change.

In the absence of readily intelligible overall information on Moricambe Bay, it is tempting to look further south to the Fylde and West Lancashire. The development of the Wyre Estuary and the Fylde, and the extent of the Mosses attested by Speed in 1610 suggests that we can expect to discover only highly localised ancient settlement (fig. 39). This is suggested by partial aerial cover obtained along the banks of the Wyre. Apart from this, preliminary results suggest that the settlement was limited to a few drumlin summits now swathed in post-glacial estuarine and marine sediments.

Further south, however, the pattern of ancient settlement in relation to coastlines, together with dune and moss systems, is arguably clearer. The pattern involved in the development of the latter is relatively easy to establish. Extensive areas of reclaimed moss formed behind coastal dunes in the valley of the River Alt and extended northwards to the Downholland mosses behind Southport to Mereside, Rufford and the river Douglas.

The presence of such moss area seems likely to be the main determinant of the location of ancient settlement. The reason for this can be best understood in relation to a food-gathering economy. The availability of food supply from alternative sources made the interface between the terminal areas of land mass and the recent lagoons, now represented by reclaimed mosses, a viable settlement zone for early man in which he could exploit alternative sources for food supply. The concept is well expressed in various current ecological publications and can be applied with advantage to the West Lancashire coast. Furthermore, Speed's map of A.D. 1610 is very useful in showing the actual presence of meres in the more waterlogged areas of what is now largely reclaimed land. The borderline of potential settlement can thus be more clearly seen and indeed is still detectable through close examination of soil maps which enable the margins of the mosses to be identified.

When extracted, the present 25–35 ft. (7.69–10.77 m) ranges represent the boundary between lagoon/moss and habitable ground, a line largely delineated by the course of the Leeds/Liverpool Canal (fig. 40). Aerial photography along this strip has for the first time succeeded in identifying traces of ancient sites in West Lancashire in the form of settlement nuclei near Downholland Cross and further north a palimpsest of field systems including a probable sub-rectangular nucleus near Heaton's Bridge. The settlement pattern continues through Burscough Bridge to the Douglas. To the north the pollen histograms produced from the Mere Sands by Dr M. Tooley strongly suggested human activity at the beginning of the proto-historic period.[8] Topographically this is matched by the occurrence of an

Fig. 38. A section showing recently identified coastal defence on the Solway, indicated by an arrow. The ditch utilises the slightly raised ground of a storm beach, showing as a bleached area between Milefortlet 3 (Pasture House) and Milecastle 4 (Herd Hill).

Fig. 39. A section of Speed's map of 1610 showing the moss areas of the southern Fylde and south-west Lancs. Note the existence of a lake at Rufford and (extreme lower left) the size of Hilbre Island.

'island' of slightly higher land projecting into the mosses and meres north-west of Rufford. Marrying the palynological with the topographical data, it was further possible to predict the occurrence of settlement and associated features in the Rufford area; settlement that, as may be seen, was in fact first located through aerial reconnaissance in 1977. Now that a major settlement indicator has been identified in this area, so long regarded as archaeologically barren, the way is open for this preliminary work to be developed into a detailed and integrated survey of sub-regional geoarchaeology.

The final sections of this discussion are concerned with two sites at the southern end of Liverpool Bay. One, Meols, at the tip of the Wirral, has long been known as a major interpretative problem; the other at Prestatyn has not been examined from the present standpoint prior to the author's work. In the case of Meols, it is relatively easy to establish that there has been substantial change at the tip of the Wirral peninsular in the last few centuries and, by extension of the argument, still earlier. Lloyd's map of Wales from 1593, for instance, shows, if it can be believed, a substantial mushroom shape landmass projecting from the tip of the Wirral.

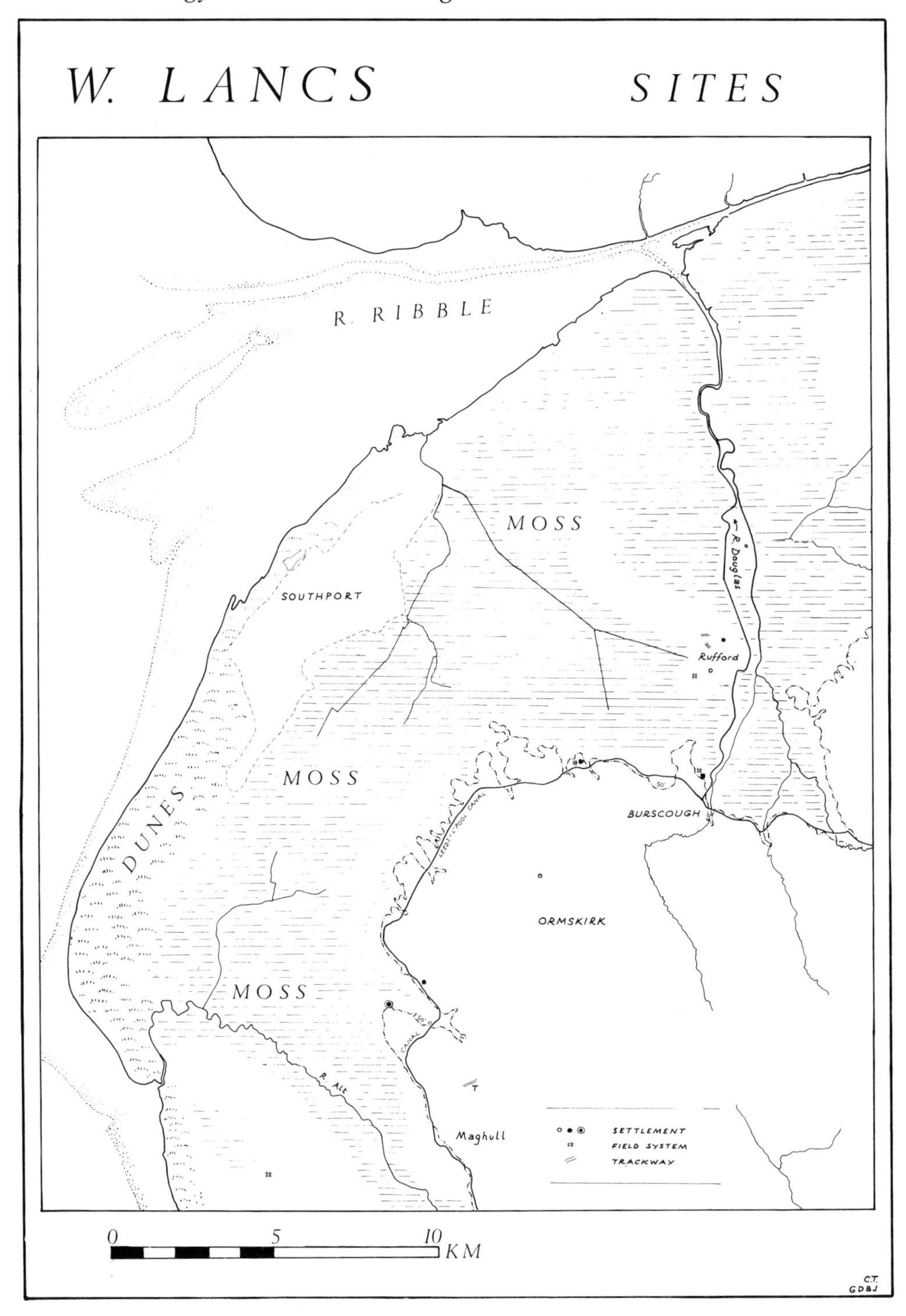


Fig. 40. South-west Lancashire: this map shows the recently established existence of farmsteads and other sites in close proximity to the 35–40 ft. (10.7–12.2 m) contour roughly indicated by the Leeds/Liverpool canal.

Coastal maps of the seventeenth and eighteenth centuries leave no doubt that the coastline has undergone substantial change in the period prior to the even more drastic alterations created by the drainage of the Dee in 1737. It is clear from the early maps, however, that a sheltered anchorage, now roughly equatable with the area still known as Hoyle Lake, existed at the tip of the Wirral in a position partly protected by Hilbre Island (fig. 41). The earliest reliable map, that of Grenville Collins in 1687 (fig. 41), shows an extensive series of sandbanks seaward of Mockbeggar Wharfe and a sand spit known as Dove Point separating them from Hoyle Sands. Dove Point has long been washed away and had all but disappeared when the account of ancient Meols appeared in 1863,[9] but its full extent may be reflected in the earliest map of 1593. Whatever the details of the evolution, the Point seems to have been directly associated with the now largely silted area of Hoyle Lake (fig. 42). Grenville Collins' chart of 1687 (fig. 41) shows that off Hoylake the channel was half a mile wide with depths of five and nine metres at eastern and western ends, whereas today it is dry at low water. It was this channel, then, that provided the known seventeenth-century anchorage still shown on the more accurate chart of 1734 and must be the basis for any reconstruction of the site's ancient form.[10] The edge of the spit at Dove Point, projecting alongside the channel would have served as a natural focus for settlement associated with the anchorage facilities. Literally thousands of archaeological finds have been made in this area but still defy generally agreed analysis.

What can be made of the evidence? The great majority of the Roman and later finds derive from the foreshore, approximately 4 km (2.5 miles) from the western angle of the Wirral. More specifically, the occupation levels that were fortunately fully recorded were found exposed in one of three land surfaces identified by erosion in the Dove Point area. The surfaces were separated by layers of silt that have been presumed to indicate transgression of one kind or another to colonisation by sand dunes before re-exposure largely in the nineteenth century. Roman finds appear to have derived from the middle land surface but no absolute chronology is available for these features from carbon dating. What appears to be suggested by the location of the various Roman and post-Roman finds is a drift of settlement westward along the coastline presumably reflecting the vicissitudes of erosion, silting and inundation. For example, most of the Saxon and Hiberno-Norse finds appear to have been made approximately ½ km (⅓ mile) west of most of those of the Roman period. It is about the Roman phase, however, that most is known. By the time of the 1863 survey approximately 3000 finds had been made along the foreshore. Indeed there is suggestive evidence to point to pre-Roman activity in the form of three Iron Age coins and even three drachmae. They suggest that Meols began life early as a beach-head trading post, just as the Saxon material enables the site to be thought of as a beach market. There are parallels between this kind of material at Meols and comparable finds from Lambay and Ireland's Eye, sites that may be interpreted as trading stations off the Dublin coast.[11] The volume of Roman material, however, begs other questions. There is very little that is non-metallic. Mainly deriving from the area east of Dove Point are numerous fibulae, spanning the first to third centuries A.D., while the known coin series extends to the end of the fourth. The preponderance of other metal artifacts, notably made in lead, and sometimes found in what were evidently traces of wattle and daub structures, inevitably raises the question of links with Halkyn Mountain,

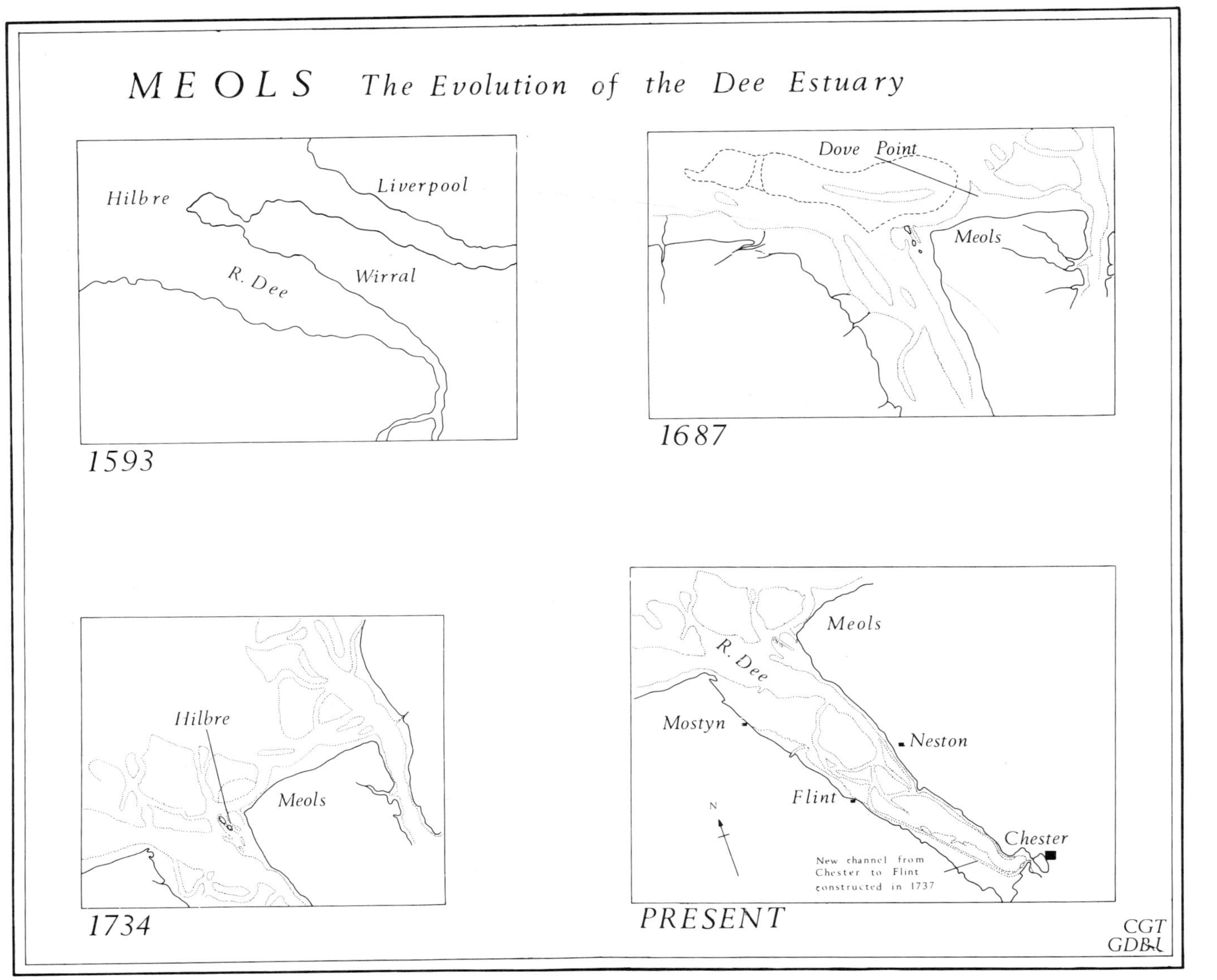


Fig. 41. The development of the Dee Estuary, based partly on Grenville Collins' map (1687) and a survey by John Eyes (1734, revised 1764), together with later Admiralty Charts etc.

Fig. 42. An aerial photograph showing the currently silted channel of the Hoyle Lake off Dove Point, Meols, looking north-east.

the nearest available source of lead across the Dee Estuary.[12] It is known, for instance, that the lead industry, as known from the associated settlement at Pentre, Flint, appears to have run down by the early third century. The extended date of occupation attested at Meols might, however, suggest that it took on the role of entrepôt for the Halkyn lead industry. A number of reasons are possible. The harbour at Pentre could never have been very satisfactory for larger vessels and material for trade might well have been better exported from the tip of the Wirral. Likewise, the extensive denudation of woodland cover caused by the requirements of smelting furnaces may have meant that by the third century it was more practical to transport dressed ore for processing across to the Wirral with its abundance of woodland cover. Certainly it seems significant that some of the latest third- and fourth-century material is firmly known to have been recovered from Hilbre Island proper (fig. 43).

It is clear therefore that Meols was much more than a fishing community throughout its long life. It is a site that cries out for an integrated research programme.

Likewise, across the Dee there is room for considerable further work on the archaeological site in the suburb of Meliden, immediately west of Prestatyn. In this case, however, recent work by the author suggests a clearer analysis of the geomorphological changes that have occurred than is currently possible at Meols.

Fig. 43. Hilbre Island looking towards the Wirral and the silted channel of the Hoyle Lake. *(Photograph reproduced here and on cover by courtesy of Airviews (Manchester) Ltd, Manchester Airport)*

Archaeologists have centred their attention on the well-known problem of the location of Varae along the Roman road into north Wales. St Asaph and, more recently, Rhuddlan have been suggested as Roman sites in this context. There is late Roman material from Rhuddlan and a persistent scatter of finds from St Asaph, the likelier of the two candidates. Whatever the truth may be, however, the largest known source of Roman finds in the area derives from Prestatyn. Excavation in the 1930s at the eastern side of St Chad's School produced the remains of two hypocausted buildings which did not appear to fit into any co-ordinated plan.[14] Accordingly, the character and development of the site has remained a matter for debate although its position must always have related to the undoubtedly ancient leadworking traceable on Graig Fawr nearby. The quantity of Roman material consistently found in the area around the school, including stamped tiles of the 20th Legion, has also always made it likely that the site was of military origin. This point was demonstrated in the extreme drought of 1976 when aerial photographs of the northern tip of the ridge on which St Chad's stands showed evidence for the northern and western sides of the double-ditched enclosure of rectangular shape. This is best regarded as the remains of an early fort from which the site developed, or, on a minimal view, the annexe to such a fort.[15] As the air photographs showed, the site was defended by a double-ditch system, of which the inner contained two sumps fronting a substantial clay rampart that had been dismantled and back-filled into the inner ditch. With this new information it is possible to re-interpret the bulk of the finds previously located to the south-east as belonging to the extensive civilian settlement depending on the continued extraction of lead into the second and probably, third centuries at the northern end of the Halkyn system.

The location of this enclosure, at the lowest point of the spur adjacent to what is a reclaimed marsh, has a special geomorphological interest for the development of the coastline. The amount of coastal accretion from the west-east silting along the present coastline has long remained a question that hindered the understanding of ancient settlement distribution. The recent work of geographers, notably Tooley, has suggested that two millennia ago the north Wales coast along with Lancashire and Wirral coastlines had a tidal limit several metres above the present high water mark. In the Prestatyn/Rhyl littoral extensive marshes that have formed behind the advancing coastal dunes have only been drained in the last century. Indeed the earliest version of the Ordnance Survey map, prepared in 1853, indicates that the present sea-front and holiday camp lay well within the tidal zone. Against this background, the siting of a fort at Prestatyn associated with the lead mining industry can probably only make sense if access to the sea was possible at the time. The present gap of some 2–2.5 km (1.24–1.55 miles) may seem large, even allowing for the time-span involved, but a glance at the unreclaimed intervening land in the neighbourhood of Prestatyn Station will show that the area concerned was, until recently, salt-marsh. Again the earliest O.S. map indicates the presence of a channel, known as Prestatyn Gutter and running E–W from the area of the fort to the coastal dunes east of the town. Furthermore, even after land reclamation, an extensive pool is marked on both the earliest and more recent O.S. maps in the area close to St Chad's School. The lowest ground level within the fort can be calculated at 4.8 m (16 ft.) above present sea-level. Immediately to the south a modern canalised stream emerges from a small valley behind Graig Fawr to run into the reclaimed marsh and it is there, amongst mushrooming estates of bungalows, that a harbourage could be postulated, easily within the tidal range of the period as newly established by the geographers. Indeed large timbers, presumably the piling of the quay, were recorded from this area early this century and are noted in the O.S. record cards. With the evidence from Prestatyn therefore, the evolution of the ancient coastline can now be better understood in an area that has always suffered severe silting from the mouth of the Clwyd.

Indeed, it is perhaps possible to take the story a stage further. The medieval motte on the east side of Prestatyn town stands beside the reclaimed line of the same Prestatyn Gutter. Air photography has again shown that the motte was larger than previously believed with an extensive bailey to one side.[16] As in the case of the Roman settlement it seems unlikely that this site can have existed without communication towards the sea. If envisaged therefore as lying effectively at the high tidal limit in the twelfth and thirteenth century, the motte gives us an approximation of the land accretion that had taken place between the Roman and medieval period (see fig. 44).

Fig. 44. Sandbanks built up by the west-east drift along the shoreline east of Prestatyn, as seen from the air. A marks the site of the Roman settlement, B that of the medieval motte.

The example of Prestatyn is important because it is one of the few archaeological sites in which one can discuss the finite remains of the settlement in relation to tidal range and various marine transgressions that have recently been postulated by geographers. Elsewhere, particularly to the north in the Fylde, fresh thinking is clearly desirable. In Lytham, for instance, marine sedimentation is recorded to an altitude of +5.4 m O.D. (*c.* 18 ft.).[17] The period of this transgression starting from

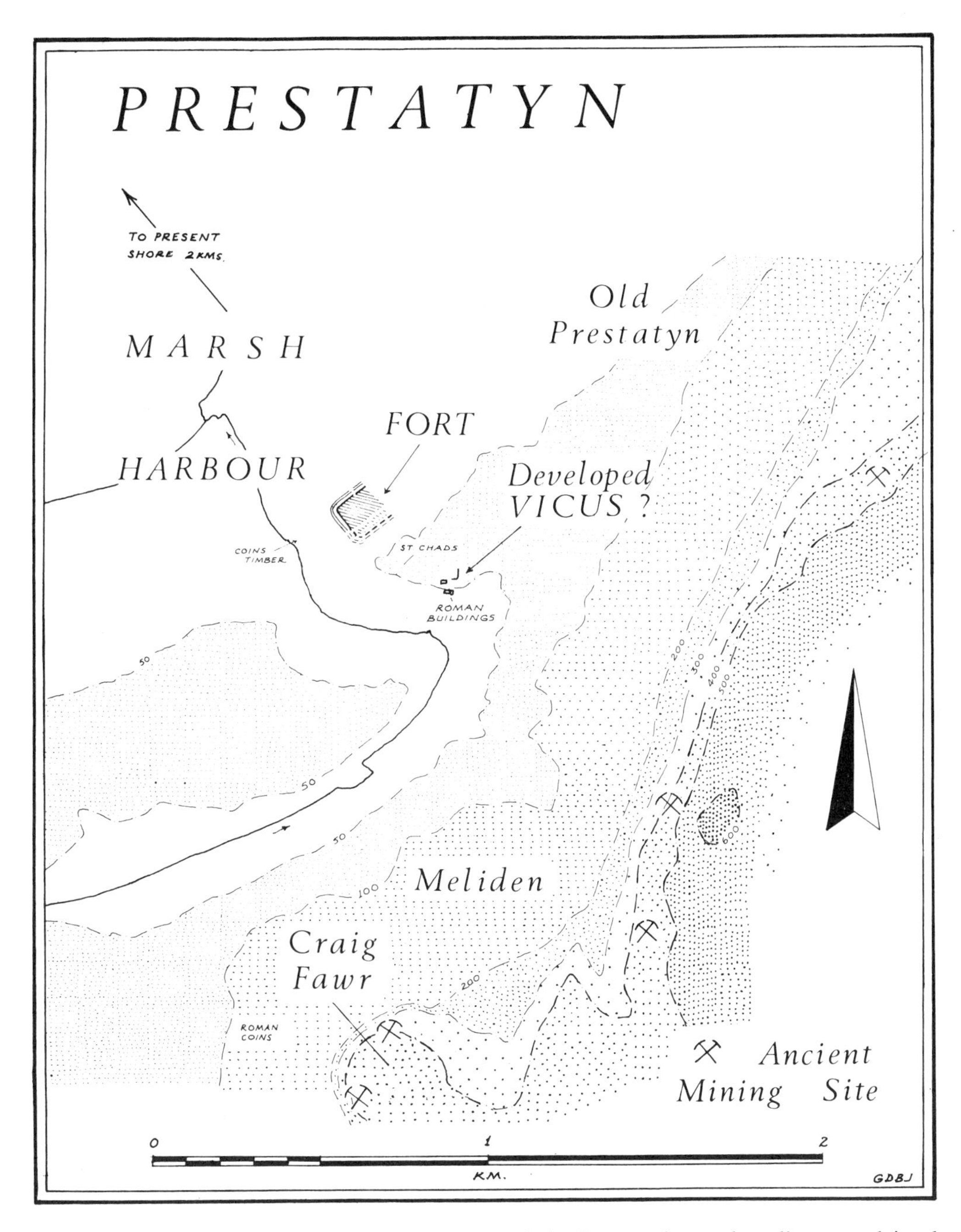


Fig. 45. Prestatyn: plan showing the relationship of the Roman site to the adjacent reclaimed marshland.

1975 ± 200 bp accords reasonably with the postulated transgression of the Fenlands and the Somerset levels. With the mean high water mark along the Fylde coast up to five metres above its present equivalent some 1,700 years ago, fresh questions need to be asked of the predictable Roman site at Skipool/Poulton on the river Wyre and at Ribchester on the Ribble. The answers can best be reached through an interdisciplinary approach by archaeologists and geographers alike.

## REFERENCES

[1] This paper owes much to consultation over the last few years with Dr M. J. Tooley, W. Cunningham, R. Bellhouse, Dr. R. Johnson and Professor D. M. McDowell. I am grateful to them and to others who made suggestions at the Day School on Coastal Change held at Manchester in November 1977.

[2] In 1882 oak trees, for example, were noted beneath the present blanket peat at Benthall, East Kilbride and East Thripland, Lanarkshire.

[3] 4150–3550 bp.

[4] H. Nichols, Vegetational change, shoreline displacement and the human factor in the late Quaternary history of south-west Scotland. *Trans. R. Soc. Edinburgh,* **67,** 145 ff.; W. G. Jardine and A. Morrison, The archaeological significance of Holocene coastal deposits in south-western Scotland. *Geoarchaeology,* **197,** 175 ff.

[5] S. Piggott, *Prehistoric Cultures of the British Isles,* 295–9; T. G. Powell, F. Oldfield and J. X. W. P. Corcoran, Excavations in Zone VII peat at Storrs Moss, Lancs. *P.P.S.,* **37** (1971), 112 ff.

[6] G. D. B. Jones, *Britannia,* **7** (1976), 236 ff.; N. J. Highman and G. D. B. Jones, Frontier, forts and farms. *Arch. J.,* **132** (1975), 20 ff.

[7] R. L. Bellhouse, Moricambe in Roman times and Roman sites along the Cumberland coast. *CW* **62** (1962), 56–72.

[8] M. J. Tooley, Sea level changes during the last 9000 years in north-west England. *Geog. Journ.,* **140,** 18–42; see now also this volume, 74–86. R. H. Johnson, A reconnaissance survey of some river terraces in parts of the Mersey and Weaver catchments. *Mem. and Proc. Manchester Lit. and Phil. Soc.* 1969–70.

[9] A. Hume, *Ancient Meols,* 1863, passim; cf. J. A. Steers, *The Coastline of England and Wales,* 1964, 104 ff. and F. H. Thompson, *Roman Cheshire,* 1965, 97 ff.

[10] Grenville Collins, *Great Britain Coasting Pilot,* 1st ed. 1893 (Admiralty Hydrography Dept. B 892); and J. Eyes, *A Chart of Liverpool Bay,* 1764 (British Museum, maps 7D.2).

[11] The first point has been made by Prof. M. Dolley, the second by Dr D. Hill; see his article, Saxon beach markets, forthcoming.

[12] For excavation reports by Taylor and Atkinson, see *Flintshire Historical Society Publications,* **9,** 58 and subsequent years; also J. Ellis-Davies, *Prehistoric and Roman Remains of Flintshire.* Recent excavations of a villa-like structure west of the apparent centre of the settlement took place in 1976–7 under the aegis of the Clwyd-Powys Archaeological Trust.

[13] Antonine Itinerary 482. 5–8; cf. A. L. F. Rivet, *Britannia* **1,** 54–5, together with Ellis Davies *opt.cit.,* 332 and W. J. Hemp, *Antiq. Journ.* **3** (1923), 69 ff.

[14] R. Newstead, The Roman Station, Prestatyn. *Arch. Camb.,* **92** (1937) 208 ff.; for the geological background see E. Neaverson, Recent observations on the post-glacial peat beds around Rhyl and Prestatyn. *Proc. Geol. Soc.,* **17,** i (1936), 45 ff. and cf. R. Newstead and E. Neaverson, Post-glacial deposits of the Roman site at Prestatyn, Flints. *ibid.* **17,** 3, 243 ff.

[15] *Britannia* **8.**

[16] G. D. B. Jones, *North Wales from the Air,* 12.

[17] See further p. 91; for further background on the 'Hillhouse coastline' see R. H. Johnson, *op. cit.* in n. 8.

# Possible Evidence for Sea-Level Change in the Somerset Levels

Dr F. A. Hibbert

The area of low-lying land, known as the Somerset Levels, has been under the influence of sea-levels since organic deposition began there some five and a half thousand years ago. Whilst deposits of a clearly marine origin, such as clay-containing marine fossils, are evident in some areas, the inland deposits, being principally biogenic in origin, may contain evidence of fluctuating sea-levels.

Recent research has been directed to the production of closely-dated peat profiles with a view to establishing absolute pollen counts for several sites within the areas of Westhay, Burtle and Shapwick Heath (Beckett and Hibbert 1979). The earliest deposits of peat were formed in a freshwater fen environment. Impeded drainage from the Levels led to a back up of ground water flowing into the area from the surrounding high land. This was base-rich in status due to the nature of the underlying rocks of this high ground. The development of the flora of the fen was characterised by species such as *Cladium* sp. (saw-toothed sedge) and other mosses, plants which demand a high base status in their water supply. As equilibrium was reached between the ground water table and the drainage pattern of the area, fen-woodland developed over the Somerset Levels and this would have remained the typical vegetation of the area had other influences not been effective. The most significant of these was high rainfall. There then followed the growth of a raised bog complex of quite different character, since it contained different plant associations, and was maintained above the ground water level due to the relatively high rainfall. Rain water is very deficient in plant nutrients and all the former dominant plant species disappeared from the fossil record.

Until the development of such deposits one might say that the fen-wood was developed at—or close to—the effective sea-level at that point in time. Kidson and Heyworth (1976) have used such levels and related them to contemporary Ordnance Datum Levels; in addition they use trackway dates and together produce a

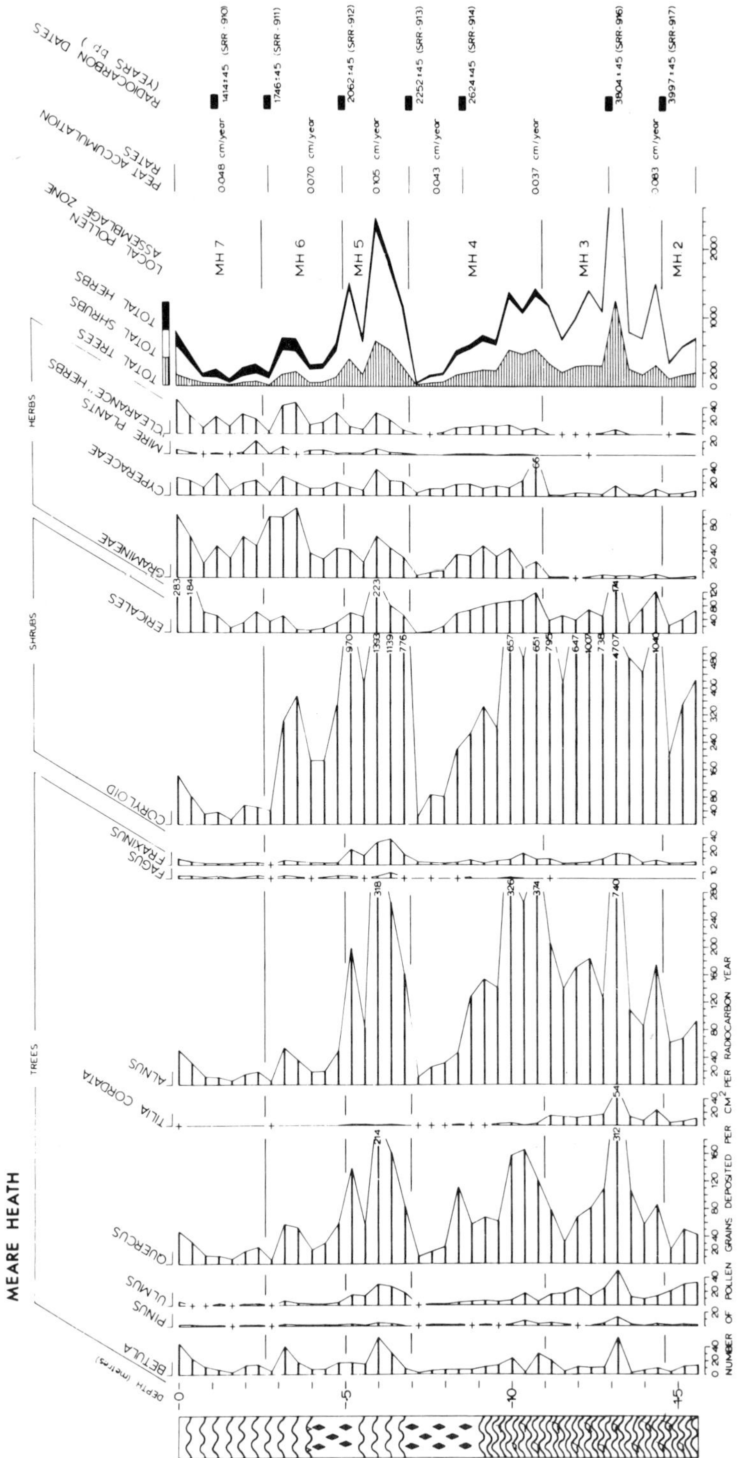


Fig. 46. Absolute pollen deposition diagram.

smooth curve for the post-glacial sea-level rise in the area. Subsequent work (Beckett and Hibbert 1979) has produced a large number of more closely related radiocarbon dates which show there to be local variation of the time that major shifts in deposition occurred within the area. This would throw doubt on the validity of selecting dates from archaeological evidence at random.

The raised bog deposits have no direct evidence of marine deposition within them. Signs of increased wetness are shown by a light coloured peat containing well-preserved plant remains. There are, however, depositions of well-preserved peat containing base-rich fen species within the raised bog itself. This indicates a substantial rise in the ground water level—sufficient to flood out the previously higher raised bog deposits and turn an acid, rain-fed ecosystem into one maintained by ground water. This must reflect a substantial change in the drainage pattern of the Levels, and if such fen wood deposits form at an effective sea level then one may interpret this reversal in deposition to reflect a higher sea level.

The pollen diagram (fig. 46), prepared by Dr Beckett, shows changes in sedimentation type over two periods, approximately from 0.4 to 0.55 m (*c.* 1–2 ft.) and from 0.68 to 0.91 m (*c.* 2–3 ft.). During these periods plants characteristic of base-rich fen communities replaced the raised bog deposits and thus reflect the change in water levels necessary to bring about such changes. The times of these changes are dated in the Meare Heath diagram. The earliest was at 2624 ± 45 bp (SRR–914) and appeared to last for a period of 150 years. The levelled heights in relation to present O.D. are +2.60 m O.D. for the onset of the earliest flooding horizon and +2.95 m O.D. for the onset of the record. By interpolation of dates the fen wood/raised bog transition at the close of the earlier flooding horizon was taking place at about 2252 bp at 2.80 m O.D. the second at 1850 bp at +3.10 m O.D.

The scale of change from raised bog peat to a ground water fen peat is too great to be accounted for simply as a change in local topography within the Levels. The changes must have been due to features affecting the main drainage pattern to the seaward end. It is suggested that this change was one of the sea-level, being of the magnitude to account for persistent flooding of the whole area. There are time correlations for these events with other areas throughout the U.K. (Tooley 1974, Godwin 1966, Willis 1961). It would seem as if the sequence of events in Somerset may be related to these and furthermore that the oscillations in changing within the raised bog sediments, back to fen, would suggest that a smooth curve of rising sea level may be too superficial a view.

## BIBLIOGRAPHY


Beckett, S. C. and Hibbert, F. A. 1979. Vegetational change and the influence of prehistoric man in the Somerset Levels. *New Phytol.* **83,** 577–600.

Godwin, H. 1966. Introductory address. In Sawyer, J. S. *et al.* (eds.), *World Climate from 8.000–0* B.C. Royal Meteorological Society.

Kidson, C. and Heyworth, A. 1976. The Quaternary deposits of the Somerset Levels. *Q. Journ. Engineering Geol.,* **9,** 217–35.

Tooley, J. J. 1974. Sea-level changes during the last 9000 years in north-west England. *Geog. Journ.,* **140,** 18–42.

Willis, E. H. 1961. Marine transgression sequences in the English Fenlands. *Ann. N.Y. Acad. Sci.,* **95,** 368–76.

# Archaeology and Coastal Change in the Netherlands

Dr L. P. Louwe Kooijmans

## *I. Introduction*

The archaeology and coastal change of the Netherlands are best seen from an ecological point of view. Governed by the rise of sea-level, sediment was laid upon sediment and a vast stratigraphy originated, in which all environmental changes are documented: a sequence of changing landscape-patterns. A major framework of this sequence is embodied in cyclic processes, known as transgression/regression cycles. The balance between land and sea was now in favour of the sea, a transgression phase during which estuarine creek systems were gradually extended and tidal flat areas were enlarged. This was followed by periodical sedimentation, during which creek systems were silted up and tidal flats changed into salt marshes. Mostly this sedimentation phase is included in the transgression part of the cycle, but it represents, in effect, the first part of a regression phase, which culminated in widespread peat formation. The transgression/regression cycle is a major topic in the present study: to what extent is this cyclicity periodical (with regular intervals) or aperiodical (with more or less random intervals)? Are there other processes that show a similar cyclicity, such as fluviatile sedimentation or dune formation? What are the agencies governing these cyclicites, and a practical consideration: are predictions for the near future possible?

In this study the earth-sciences and palaeobiological disciplines combine forces: the vegetation is reconstructed by a palaeobotanist with the help of pollen, while macrofossils (seeds, wood), molluscs, foraminifera and diatoms help to reconstruct water conditions (tidal ranges, stream velocity, presence of mud and salt). Sometimes skeletal material of fishes and mammals permit a glimpse of the macrofauna.

The chronological framework is formed, together with stratigraphy, by many hundreds of C14 dates of a high standard, all determined by the Groningen Isotope Physics Laboratory.

## *II. Geological Differentiation of the Delta*

The area under discussion is not homogeneous at all, but comprises widely different landscapes. I only want to give here a short sketch and refer for more detail to my earlier description (Louwe Kooijmans 1974) and the literature listed below. Better than words can do, the map and section illustrate the way the area is built up of different types of sediment, under the influence of the rising sea-level (figs. 47, 48).

Deposition started with a basal peat and was followed by tidal flat and salt marsh deposits behind coastal barriers as the sea encroached farther. East of the marine and estuarine environments the peat growth continued and behind this zone a fluviatile sedimentation district was established along the main rivers. A major change took place around 3000 bc, when the gradual eastward shift of the coastal barriers came to an end. New coastal barriers were subsequently formed seaward of the old ones. The 'intracoastal zone' (i.e. the land between the coastal barriers and the Pleistocene hinterland) became better protected from marine incursions and changed into an extensive freshwater swamp. Only behind the inlets through the barriers did marine and estuarine deposits of more restricted extent form during 'transgression phases'. In relatively recent (i.e. post-Roman) times considerable destruction of the land and renewed marine deposition took place in those parts, where the coastal barriers extended the farthest to the west:

in the South-West, the province of Zeeland and adjacent regions
in the North, the northern parts of Holland and in the IJssel Basin behind.

In these regions remnants of the old deposits and traces of occupation in the old landscapes have only accidentally survived.

In the northern Netherlands (the provinces of Friesland and Groningen) the conditions were rather different. No coastal barrier remains are known there and from the facies of the intracoastal deposits it can be concluded that throughout the Holocene the coastline must have been interrupted with inlets, as nowadays. Deposition took place mainly as tidal flats and salt marshes. Peat formation took place in only a small belt and no major rivers flowed to this part of the coast. The coastal district is made up of salt marsh deposits from 600 bc and later, behind the present islands and Wadden Sea.

So we can distinguish a number of zones between the actual coast line and the hinterland, characterised by decreasing marine and increasing freshwater influences:

1. The sandy coastal barriers, covered by dunes
2. The tidal flats
3. The salt marshes
4. Estuarine creek systems
5. Peat zone
6. Districts with fluviatile sedimentation.

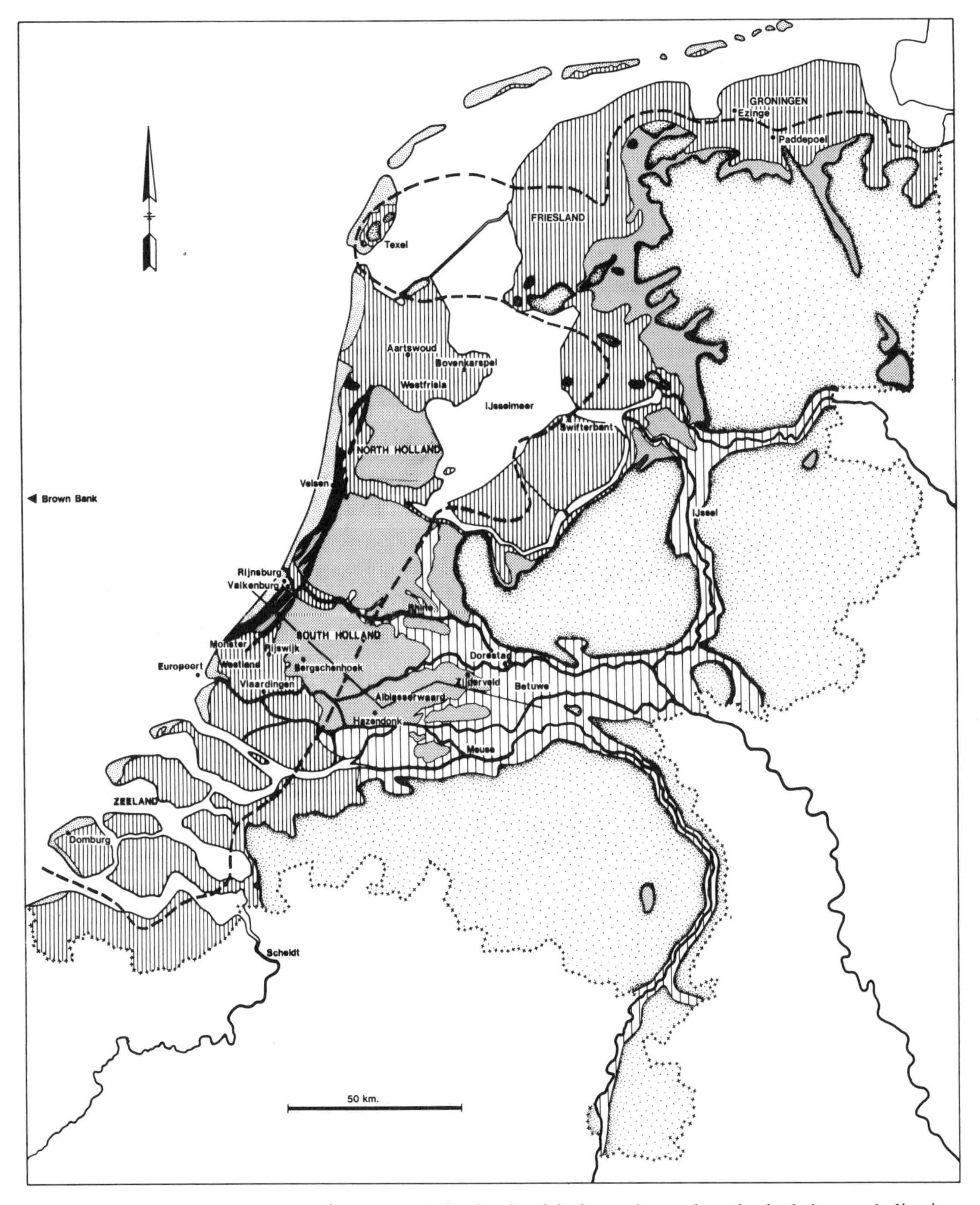


Fig. 47. Generalised geological map of the Netherlands with the major archaeological sites and districts mentioned in the text. Geology after S. Jelgersma *et al*. 1970.

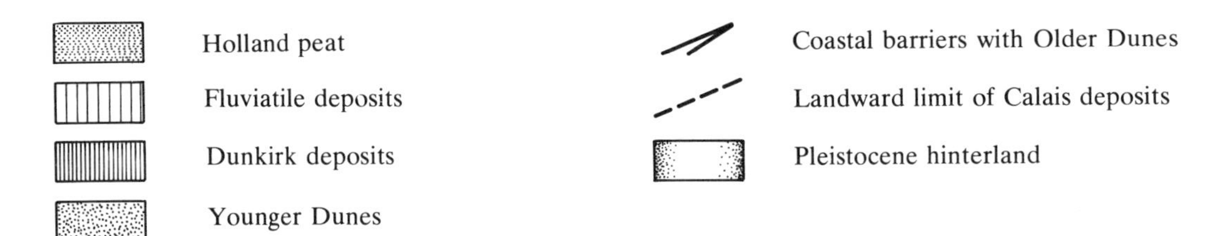

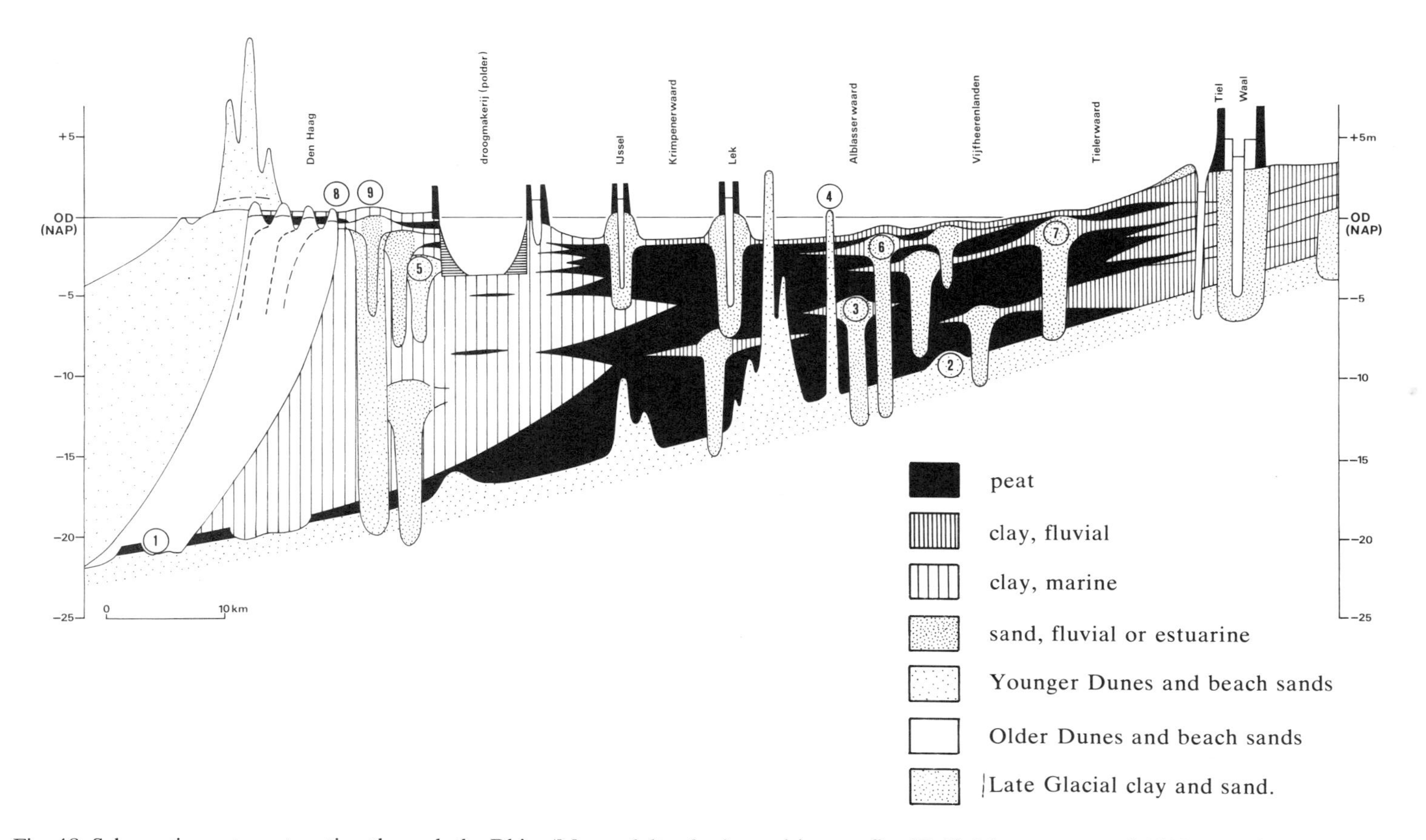


Fig. 48. Schematic west–east section through the Rhine/Meuse delta; for its position see fig. 47. Height exaggerated 1400×. Indicated are a number of sites (or their projection onto this section) of which the present levels were used in the construction of the sea-level curve of fig. 50.

*Indicated sites*:
1. Europoort 2. Willemstad 3. Swifterbant S3 (projected) 4. Hazendonk 5. Vlaardingen, Hekelingen 6. Molenaarsgraaf 7. Zijderveld 8. Voorschoten, Loosduinen, Arentsburg 9. Valkenburg, Rijswijk

The position and width of these zones fluctuated over time and their widths show considerable variations along the coastline. As regards these variations the intracoastal district can be divided into a number of sections from south to north:

1. The estuary of Scheldt and Meuse
2. The Holland plateau
3. The Lake IJssel basin
4. The northern salt marshes.

The river clay area may be distinguished as a fifth district.

Although one is inclined to speak of the combined 'delta' of a number of rivers, especially the Scheldt, Meuse, Rhine and Vecht, there are in fact only some minor parts in the area as a whole where true delta deposits were laid down. The 'Rhine/Meuse delta' consists of a complex of marine, estuarine, fluviatile and organic deposits.

References: Hageman 1969; Jelgersma *et al*. 1970; Jong 1967; Jong 1971; Pons *et al*. 1963; Roeleveld 1974.

### *III. Human Occupation: Where and When?*

In which areas did people settle in these landscapes before the embankments were constructed in the Middle Ages (from *c*. A.D. 1000)? To what extent was this choice of terrain governed by their subsistence economy and/or to what extent did they adapt themselves in this respect?

(a) On the coastal barriers and the Older Dunes, settlement traces are found from the Vlaardingen culture (2400 B.C.) onwards. No occupation is found from barriers at the time they formed the actual coastline. The settlements are situated on the older ones more inland. There are no gaps of considerable importance in the occupation sequence.
(b) No remains are known, nor are they to be expected, from the former tidal flat regions.
(c) On salt marshes and comparable deposits, the slightly higher and more sandy parts were chosen, like surf ridges, creek levees and completely silted-up creeks. In Westfrisia extensive occupation is dated 1200–700 B.C., in Groningen and Friesland from 600 B.C. onward.
(d) In the estuarine districts people founded their settlements on the silty levees of tidal creeks or on the sand-bodies of those that were silted up.
(e) In the peat districts occupation did not take place on the peat surface until the general reclamation of the eleventh–twelfth century A.D., with the exception of some native settlements of the Roman period. But where sandy stream ridges and outcropping dune-tops were available, i.e. in the extension of the fluviatile clay districts of IJssel and Rhine/Meuse, these were settled throughout prehistory, from the Mesolithic to the Iron Age.
(f) In the river clay area the levees of active (and perhaps also former) rivers were occupied from the Late Iron Age till recent times. Older (Late Neolithic and Bronze Age) settlements were found on older systems and especially on so-called crevasse-deposits. In general, these are the well-

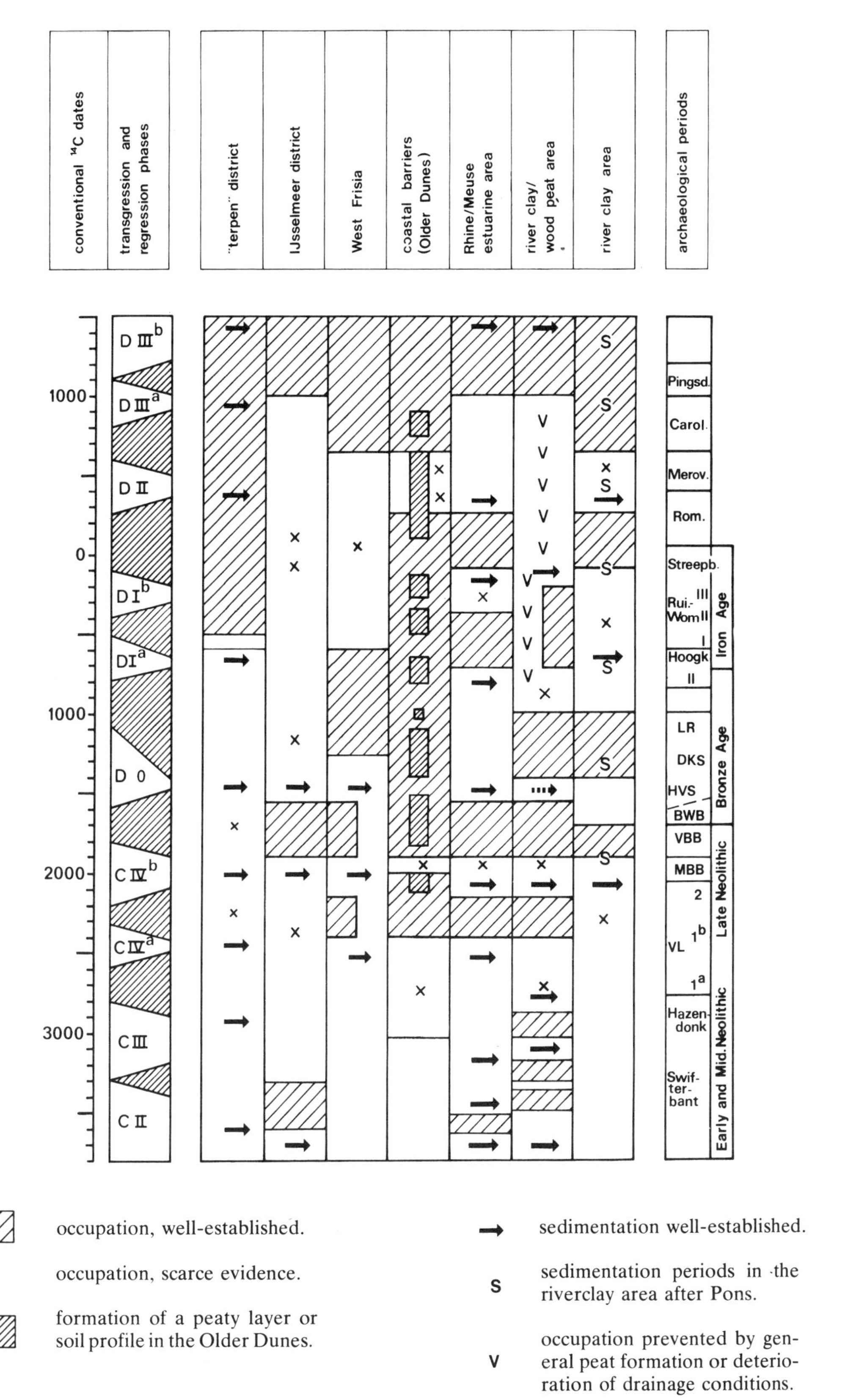


Fig. 49. Schematic representation of the occurrence of settlement traces in the various physiographic landscapes of the Rhine/Meuse delta between 4000 B.C. and A.D. 1500, compared with the transgression/regression-cycles. After L. P. Louwe Kooijmans 1974, with some modifications.

drained and relatively sandy deposits formed next to open watercourses of some importance, up to local MHW level or even slightly higher, and then preferably the higher parts of these. The coastal Older Dunes and the out-cropping dune-tops in the peat district form special cases, where the interaction of environment and occupation, although certainly present, is of a more restricted degree.

From this inventory the criteria that played a role in the choice of terrain for settlement can be extracted. The major factors for all communities, irrespective of the way of life, are:

(a) As little inconvenience as possible from water, which means adequate drainage and a height above the level of regular flooding.
(b) The economic possibilities of the site and its direct surroundings. Fresh water had to be available and, for rural communities, enough land for cattle and crops. Where these conditions are not fulfilled hunting, fowling, fishing and collecting played an important role, as at Swifterbant, Bergschenhoek, Hazendonk and even in the late Neolithic Vlaardingen. On these sites one must try to answer an important question: was the occupation perhaps seasonal or connected with a special, season-bound activity? This seems to be the case at Swifterbant. The sites are probably summer fishing camps. At Vlaardingen, however, permanent occupation is the most likely. The Hazendonk research is not yet far enough advanced to answer this question.

As stated above, the quantity of data has increased so much that we can safely say that landscapes were inhabited whenever and wherever possible. We must now turn to chronology and ask when and why occupation starts and ends. In the sedimentary phase of a transgression/regression cycle high levees, etc., could be formed, especially where the high waters were raised, when collateral flow was restricted, for example because of the silting-up of small side creeks during this phase. When subsequently a drainage pattern changed—and this was often the case at the end of the sedimentation phase—it could easily happen that these high deposits were outside the reach of (or at least far away from) normal high water, be it marine, estuarine or fluviatile. In this way, by the lowering of the local MHW (i.e. often a decrease of the local tidal amplitude), a favourable situation for settlement came into existence. In our opinion this was the case, for instance, with Bell Beaker occupation in the Alblasserwaard peat district, Middle and Late Bronze Age occupation in Westfrisia, Iron Age and later colonisation of the Groningen and Friesland salt marshes and 'indigenous' Roman occupation in the Zaanland peat district north of Amsterdam, in the estuary of the Meuse (Westland) and in the Betuwe river clay area.

The major cause for the ending of a period of occupation was the general rise of sea-level, which counteracted the temporary gain of the sedimentation process. In due course the favourable conditions that led people to settle down disappeared; drainage became worse, flooding more frequent, arable land and natural pasture land marshy. A new transgression phase might have accelerated this process.

This process was more pronounced in earlier prehistory, when the rise of sea-level was rapid, during the Neolithic, for instance, about 20 cm/century (8 in.); in the Roman period about 5 cm/century (2 in.). At the end of the Late Bronze

Age people of Westfrisia were the first to fight back, by retiring to the highest points, digging ditches around the farms and by raising the yards with the excavated soil. But it was not a success. Sea-level rise, or more correctly the resulting rise of ground water level, won. Slightly later the northern salt-marshes were settled and there, about 500 B.C., people managed to resist flooding by raising small artificial mounds of sods, called *terpen* or *wierden*. The *terpen* were raised when necessary and also extended after the Roman period, in order to provide land for crop farming, when the old salt marsh surface became too wet.

When the embankments and artificial drainage were constructed, between A.D. 1000 and 1300 in most of the low districts, natural conditions played only a minor role when new settlements were founded. Many of the present-day villages in Holland are of medieval origin and lie in the middle of the peat bogs. But the individual farms often had raised yards. Until recently the dykes proved to be no safe guarantee against flooding.

References: Clason 1967; Louwe Kooijmans 1974; van Regteren Altena *et al.* 1962–3; van Zeist 1968; van Zeist 1974.

## *IV. Sea-level Records*

A special topic is that of the rise and possible fluctuations of sea-level: the study of dated levels and the construction of a time-depth curve (fig. 50). It is the mechanism governing the main line of geological history, for which archaeological research may produce very reliable and extensively controlled data. The research concentrates on the collection of such data, on the estimation of the margins of error and the factors of local and regional importance that must be taken into consideration, on the establishment of variations along the coast and on the question whether and to what extent the transgression/regression cycles are reflected in the time/depth curve.

I explained in a special paper (Louwe Kooijmans 1976c) the line of reasoning that must be followed to attain a dated MSL from a dated sample-height. There are four steps in this line, each with its margins of error, that accumulate in the end-result.

The first step is from the present altitude of the sample to the original level. Sources of error are: the margin of error of the measurement itself, the relation of the sample to the presumed level and, above all, compaction. Preferably samples must be compaction-free. Otherwise compaction must be estimated, which is often impossible with any degree of accuracy. The importance of compaction-free samples cannot be overstressed.

The second step is to establish the position of a former local water level from this (present) height. The relation between the sedimentary height and a certain water level must be evaluated for this purpose. This can be done by comparison with recent situations. A salt marsh level lies at about 40 cm (15.5 in.) above MHW and estuarine creek levees are silted up to MHW level. But much basic study still must be done in this field.

The third step is that from the local water level, which mostly is the MHW, to a more general valid value; how representative is the sample-point for the whole

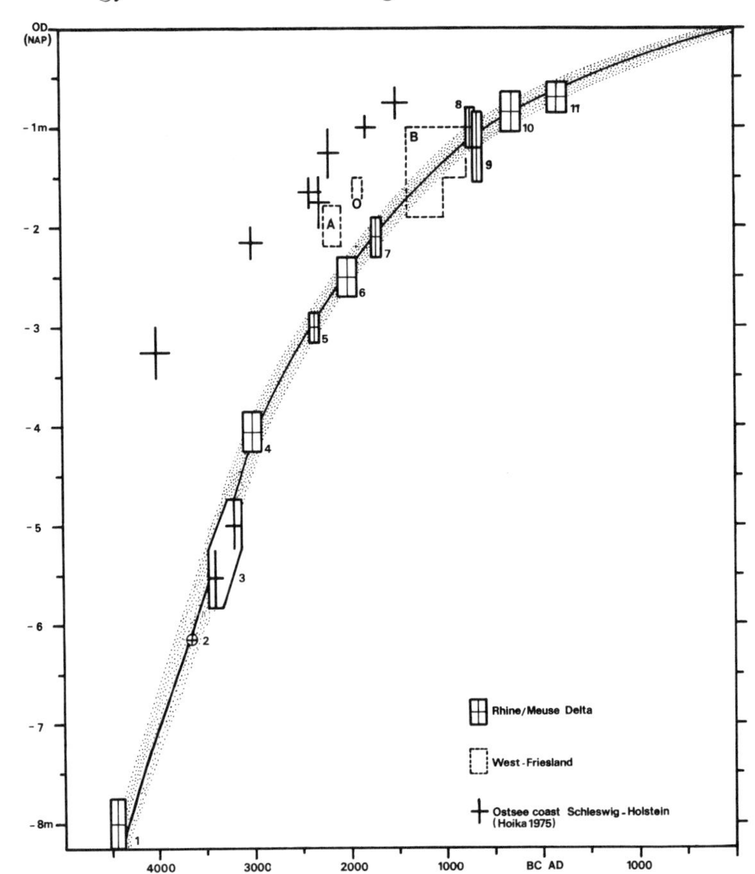


Fig. 50. Time/depth diagram of dated MHW-levels in the Rhine/Meuse delta, derived from the depths of archaeological sites, and the curve for the rise of mean sea-level that is obtained in this way. The dotted zones give an indication of the variation due to locally or regionally varying conditions.

Also indicated (but not taken into account) are some data from Westfrisia, where sedimentation during the Calais IV phase reached very high levels.

The surve of the eustatic rise of sea-level must lie somewhere between the data from the area of tectonic depression at the mouth of the river Rhine and the area of glacio-isostatic upheaval of the east coast of Schleswig-Holstein.

area? There are many factors that cause regional and local variation of MHW, namely:

(a) Extinction or enlargement of tidal amplitude by flood depression (in wide basins) or stowage (in narrow channels) respectively.
(b) Rise of all water levels when we go upstream along the lower courses of small and big rivers (gradient effect).
(c) Rise of the groundwater table in extensive sand bodies, like the coastal barriers, caused by restricted velocity of drainage.
(d) When relatively large areas are studied, like the whole of the Netherlands, MHW-variations along the coast cannot be neglected.

The fourth step is that from MHW to MSL. The former tidal amplitude must be estimated.

It will be clear that the best points for sampling are those where only a limited number of factors played a role, or situations where the effect of different factors can be compared, or thirdly, regions where the comparison of samples of which all factors are equal is possible.

I tried to construct a curve based on archaeological sites, selected according to this principle and following the line of reasoning above (Louwe Kooijmans 1976c). A much more thorough study is being made at present by van der Plassche, based on a long series of dated peat samples taken from carefully selected sites, controlled by very detailed geological mappings.

As to the results the following remarks can be made:

(a) The main path of the time/depth curve is now firmly established, with a rise of about 20 cm/century (8 in.) during the Neolithic, diminishing to about 5 cm/century (2 in.) since the Roman period.

(b) This curve gives the relative rise of sea-level. The post-Roman rise may be for the greater part the result of tectonic sinking of the land (calculated to 2–4 cm/century (0.8–1.6 in.)); the earlier rise is mainly caused by the eustatic rise of sea-level.

(c) There are minor differences between Zeeland, Holland and Groningen, caused by differences in tectonics, morphology and tidal amplitude.

(d) MHW-fluctuations established on archaeological sites, especially temporary lowering of MHW, can be explained for the greater part (if not completely) by local changes in drainage patterns. We can presume that these, running parallel to the transgression/regression cycles, are reflected in an ideal time/depth curve (Louwe Kooijmans 1974, fig. 14), but to measure these small fluctuations seems beyond our capabilities in the field. If ever possible, this should be the case in van der Plassche's investigations.

References: Bennema 1954; Jelgersma 1961; Jelgersma 1966; Louwe Kooijmans 1976c; van der Plassche (forthcoming).

## *V. Some Examples*

In the following paragraphs an introductory description will be given of recent research in some widely different regions of the area dealt with in this paper, to illustrate the diversity of the problems and information gained there.

### 1. EARLY MESOLITHIC IMPLEMENTS FROM THE NORTH SEA AND EUROPOORT, ROTTERDAM

In addition to the well-known barbed point from the Leman and Ower Banks off the Norfolk coast, more bone tools were retrieved from depths between 35 and 45 m (115–150 ft.) west of the Brown Bank in the North Sea. They give a depth of −40 ± 5 m (−131 ± 16.5 ft.) at 7000 ± 400 bc, where (in view of the preservation of bone) peat formation took place. This is in good agreement with C14 and pollen-dated moorlog samples. They offer, moreover, information about affinities between the material cultures of the Mesolithic hunter–fisher–gatherer communities of Great Britain, the Netherlands and Denmark.

Small bone implements, dredged up during harbour construction of Rotterdam-Europoort and found at the surface of the new artificial sand plain called 'Maasvlakte', give additional information in the last respect, but not about former sea-levels. The 'gradient-effect' plays a role and the substantial margin of error in the dating.

Together with modest finds on sandy outcrops (some of them covered) as at Swifterbant and Hazendonk, they document an occupation of unknown character, when peat formation started, i.e. in the 'North Sea Land' during the Early Boreal, and on the sands in the subsoil of the Western Netherlands in the later part of this period.

The rapid rise of the sea-level (*c.* 2 m/century (6.5 ft.)) had an enormous horizontal effect in the flat North Sea Basin and people living there must have been driven back by the encroaching sea. The appearance in some parts of the Netherlands in the beginning of the Atlantic of the flint assemblages of the De Leien–Wartena Complex must reflect these people settling in more inland areas. There are now more arguments for this supposition. Firstly, the material equipment of both (the Boreal and the Atlantic) groups, although very different in character, has distinct Nordic traits. Secondly, in both cases a marshy environment was preferred.

References: Louwe Kooijmans 1970–1; Newell 1973.

## 2. SWIFTERBANT — EARLY NEOLITHIC SETTLEMENTS ALONG FRESHWATER TIDAL CREEKS, *c.* 3300 B.C.

In the newly reclaimed IJsselmeer polders an estuarine creek system of Calais II age, part of the then IJssel estuary, was found preserved a little way below the present surface, which is at about −5 m O.D. (−16.5 ft.). On the levees settlement sites were discovered, dating from the final stage of the sedimentation period. These were investigated during the last years by the Biological–Archaeological Institute, Groningen.

In a fresh or only slightly brackish environment, with a very small tidal range (10–20 cm (4–8 in.)) the narrow, clayey levees were only incidentally flooded. They were covered with a deciduous forest of oak, elm and lime; in the swamps behind mainly alder brushwood was found, with willow-reed marshes further behind. The small early Neolithic settlements measured less than 30 m in cross-section and were intermittently occupied during a period of about one century. There are geological arguments for the absence of winter occupation and palaeobotanical evidence points even to non-annual (summer-) returns to one site, which might imply the alternate use of different sites by one community. But there are other arguments for longer stays.

A small cemetery has been found, cattle and pig were kept and slaughtered at the site, and Naked six-row barley and Emmer wheat were grown in the restricted space that was available. Apart from this, beaver, otter, red deer and wild boar were hunted. Future identifications of the bones of moor- and water-birds (which were also hunted) might give valuable information about the seasonal occupation

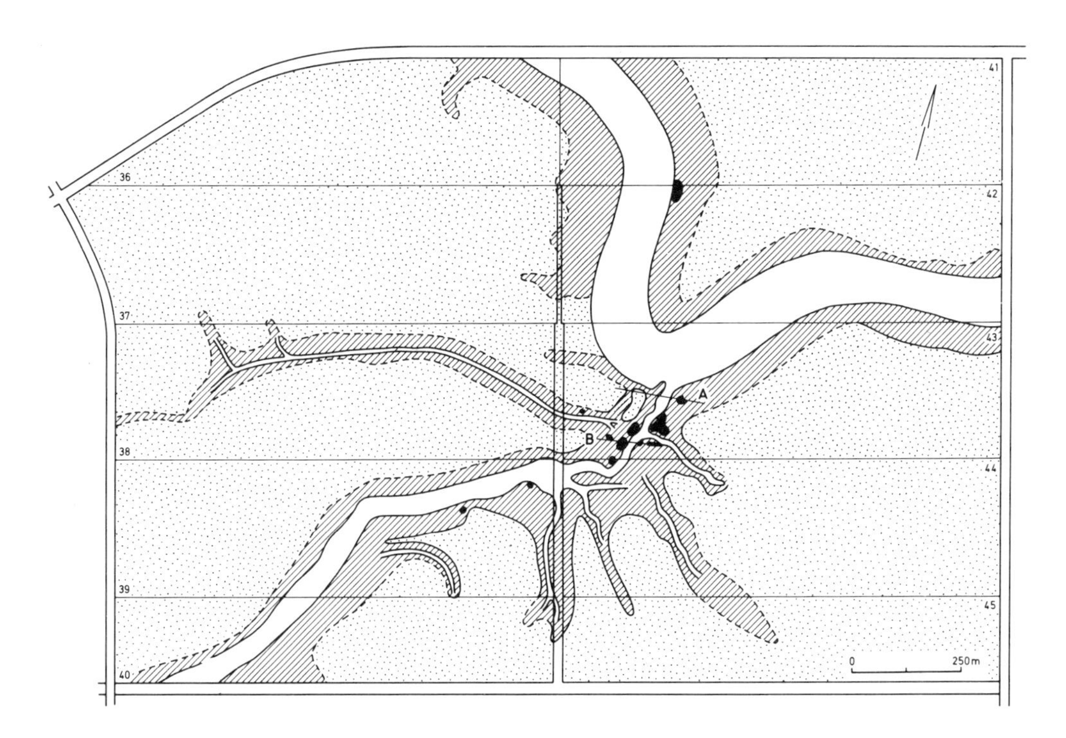


FIG. 2. — The Calais II deposits in the environment of the Swifterbant levee sites.

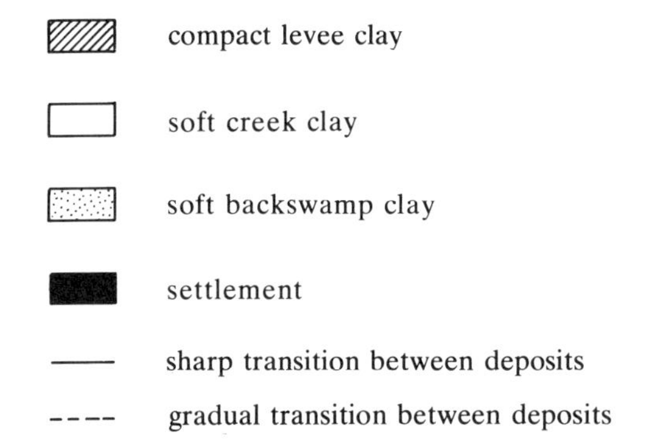


Fig. 51. Detailed palaeogeographic map of a part of a Calais II estuarine creek system with a top at −5 m O.D. (−16.5 ft.) in the new Flevoland polders and the Early Neolithic settlement sites on its banks. After L. Hacquebord, Swifterbant contribution 3, *Helinium,* **16,** 1976, 38.

of the site, by the presence or absence of migratory birds. The fruits found (hawthorn, apple, hazelnut, rose hips and blackberry) must have been collected in the late summer/early autumn, while some of the fish (especially sturgeon, salmon and grey mullet) were only present in these waters in spring and early summer. In view of the position of the sites, at a crucial point of the creek system, fishing must have been of major importance.

The material culture reflects a local evolution from Mesolithic communities, with a pottery in a Nordic (Ertebølle) style and (trade) relationships with late Rössen communities, proved by the presence of fragments of true *Breitkeile*. The sites must have been left when the creek system silted up and this area lost its special attraction. From the occupied levels an MHW at −5.55 ± 30 cm O.D. (−18.2 ± 1 ft.) around 3300 B.C. could be calculated.

To what extent are these data representative of the occupation of the total area of the then 'delta'? No coastal barriers from this period are preserved and so all possible coastal sites are lost. Major parts of the intracoastal area are eroded and the remaining parts are deeply covered. By lucky accident a few glimpses of the former occupation are permitted: at Swifterbant, where the later cover has been eroded in historical times; at the Hazendonk, a sandy out crop of only 2 ha (5 acres), where grain was grown (Einkorn and Naked barley) and large amounts of fish refuse indicate an important fishing activity; a third site was discovered recently near Bergschenhoek, north of Rotterdam, in an artificial pond, dug to −8 m O.D. (*c.* −26 ft.). This site, excavated in 1978, appeared to be a very small (winter?) fishing camp, used 5–11 times within a period of only 5–7 years as could be concluded from the microstratigraphy. Both sites lie in a freshwater peat landscape, not dissimilar to Swifterbant.

It seems that the occupation pattern of the later Mesolithic, with maintenance camps and extraction camps, persisted into the Early Neolithic, while hunting, fowling, fishing and gathering were combined with animal husbandry and crop farming. At this stage nothing can be said about the velocity and pattern of this change; was it gradual or abrupt, did it happen in stages, and are there regional variations? Informative sites are so rare, that one may wonder whether answers to these questions may ever be found.

References: Louwe Kooijmans 1976b; Swifterbant Contributions 1–8; van der Waals 1972.

### 3. HAZENDONK AND MOLENAARSGRAAF—NEOLITHIC OCCUPATION IN THE PEAT DISTRICT, 3400–1700 B.C.

Where the Late Glacial Rhine/Meuse Valley underlies the peat district, many tops of river dunes are found, dated to the Pleistocene/Holocene transition, and for the greater part submerged by the peat formation. Through the millennia they were dry islands in the marshes and ideal places for prehistoric people to settle down. The small Hazendonk, measuring only 50 × 200 m (160 × 650 ft.) and in an isolated position, had attracted people in at least nine successive phases of the Neolithic and once or twice earlier still.

This district is a freshwater, eutrophic peat area, outside tidal and marine influences, but open to fluviatile incursions from the east. There is a sequence of phases with general peat growth alternating with wide spread flooding and clay deposition along creeks and in lakes. The marshes were covered by an alder brushwood and reed, while the sand dunes had a cover of deciduous trees such as oak, elm and lime.

Traces of the successive Neolithic settlements on the top of the Hazendonk are completely lost, but old surfaces with domestic refuse on the slope of the dune and in the covering peat are preserved in excellent stratigraphy. This refuse was sampled by hand and by sieving during three excavation campaigns (1974–76), led by the author. Apart from pottery, stone and flint tools, bone implements, worked wood, charred seeds, grain and fruits, animal bones, fish remains and some human skeletal material were also collected. Pollen samples and series of C14 samples were selected. The detailed geological mapping of the complete Holocene deposits (about 10 m) of the immediate surroundings (*c.* 4 km² (*c.* 1.5 sq. miles)) of the Hazendonk is a special project to obtain palaeogeographic maps of each occupation phase. This is being undertaken by the Institute for Earth Sciences, Free University, Amsterdam. Traces of both earliest occupations of the sites were discovered in this project and were not reached in the excavation.

Although finds and samples are very unevenly distributed over the various occupation phases, due to the restricted or superfluous traces that are left of these, it will be possible to follow the changes in material culture, food economy and landscape and their mutual relationships on this site in detail over seventeen centuries, covering the entire Neolithic period. All occupation phases are separated by periods in which the disturbed vegetation had completely recovered. For some phases the character of the occupation (incidental, seasonal or permanent) and an impression of its duration can be obtained. The 'attachment-points' of the refuse layers to the sandy dune slope give very reliable sea-level data, only to be corrected for the 'gradient-effect'. It is remarkable that the sequence of occupation seems to have no distinct relationship with the sedimentation sequence of the surroundings, which is in sharp contrast to other areas and to expectation.

It would be premature to give explanations or to make estimations of the results that will be available in some years. Work is in progress on all aspects of this. The coordinates and codes of 40,000 individually mapped finds are at present being fed into the computer, drawings are being made, C14, seed-, pollen- and fish-samples are being analysed; the bone refuse is being studied by Dr A. T. Clason. But some comments can be made.

In the lowest levels (3400 B.C.) the settled area seems to be small (*c.* 500 m² (*c.* 5400 sq.ft.)). The density of finds increases from the lower layers upward. In every phase crop cultivation has been proved (pollen), but charred grain occurs mainly in the lower layers. In every period fishing was important, in view of the abundant fish remains; sturgeon fishing and hunting (especially of beaver and roe deer) were of considerable importance as late as 2100 B.C. (Late Vlaardingen/AOO Beaker).

We have no explanation why people at a given moment stopped using the site; there might be different explanations for the various periods. During the Late Vlaardingen occupation a large stream crossed the peat about 1 km (0.6 miles) north of the site. We can imagine that sturgeon was caught there and that the Hazendonk lost its attraction when this stream sanded up. The modest Late Bell

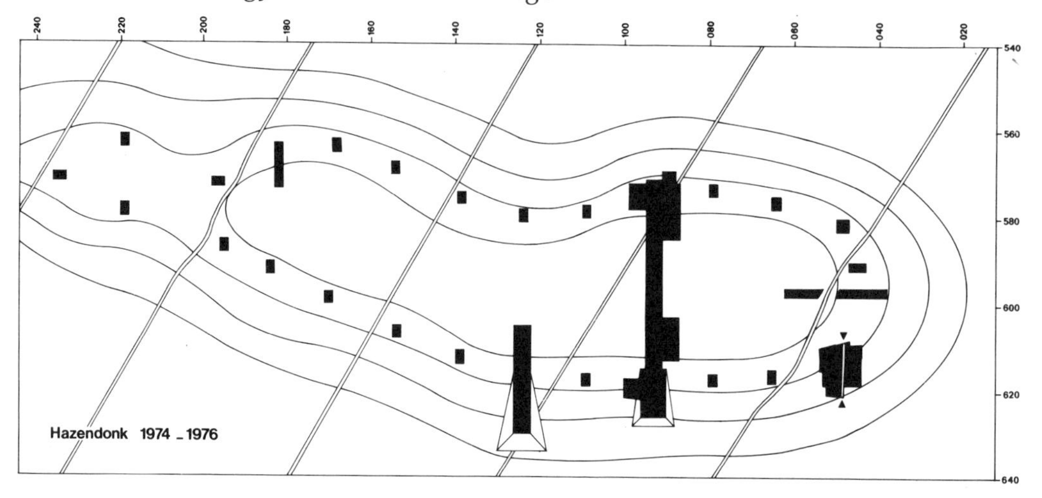


Fig. 52a. Hazendonk, generalised contour map of the surface of the sand dune with contour lines at 0, 2, 4 and 6 m (0, 6, 13, 20 ft.) below the present surface, which itself is at −1.30 m O.D. (−4.3 ft.). Excavation trenches and sampling pits in black. Grid of 20 m squares indicated along the margins.

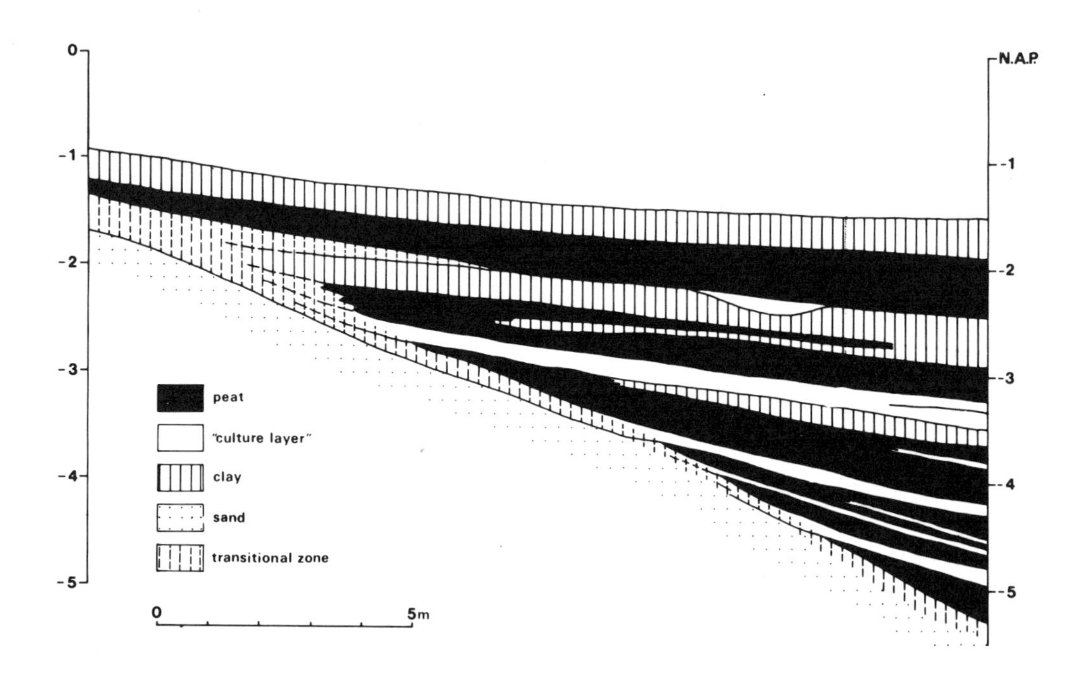


Fig. 52b. Simplified section through the deposits that cover the sand slope in the SE-trench. This is the most complete stratigraphy on one point of the site. Missing culture levels (VL2 $^{b1\ and 2}$) are projected from other sections. Height exaggerated 2×.

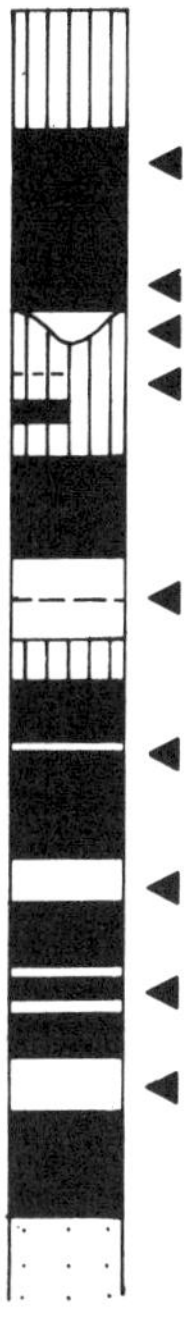

| phase | age | depth | pollen | affinities, remarks |
|---|---|---|---|---|
| | | | | |
| 12th cent. A.D. | | | 7 | |
| Late Bell Beaker | 1800-1700 | -1.90 | 5 | Molenaarsgraaf |
| Vlaardingen-2b (2) | 1900 | | - | |
| Vlaardingen-2b (1) | 2100 | -2.10 | - | Late PF & AOO Beakers, Voorschoten 2b |
| | | | | |
| Vlaardingen-1b | 2500-2400 | -2.55 | 4 | Drouwen C/D, Voorschoten 1 |
| Vlaardingen-1a | 2700 | (-3.00) | - | |
| Hazendonk-3 | 3000 | -3.50 | 3 | Het Vormer, Cuyk |
| | | (-3.65) | | |
| Hazendonk-2 (a,b) | 3200 | -3.80 | - | Belgian MK, Grimston-Lyles Hill bowls |
| Hazendonk-1 | 3400 | -4.35 | 2 | Swifterbant, Boberg 15 |
| Hazendonk-0 | c. 3700 | -5.10 | 1 | old surface(s) in borings |
| Mesolithic | c. 4500 | c.-8. | - | old surfaces in boring |

Fig. 52c. Stratigraphic column, similar to the extreme right part of fig. 52a, with archaeological comments.
Age: preliminary C14 dates bc.
Depth: local mean water-level (= sea-level + (50 ± 20 cm)).
Pollen: local pollen zone.

Beaker occupation and the absence of later refuse can be easily explained by the presence of a sand body, a few kilometres long and about 100 m wide (330 ft.), left by this stream. Intensive prospecting combined with three excavations on this so-called Schoonrewoerd Streamridge offered a detailed picture of the occupation around 1700 B.C. Over *c.* 5 km (3 miles) small sites are spread at intervals of 300–800 m (1000–2500 ft.), representing the positions of single farmsteads. Crops were grown on the deforested sand ridge, and cattle were grazed on natural pasture land on both sides.

Hunting was unimportant (enough arable land being available), but fish might still have been a prominent source of food. The long houses were spindle-shaped (if my interpretation is correct). In a temporarily abandoned settlement some burials were made.

Peat growth continued and the absence of new mineral sedimentation made the region less attractive and this became the major cause of a gradual eastward shift of the occupation. In the Roman period the region was completely abandoned and people returned only in the eleventh century when the present day villages were founded and the peat marshes were reclaimed.

References: Louwe Kooijmans 1974, 1976a, 1976b.

### 4. THE DUNE LANDSCAPE OF THE COASTAL BARRIER BELT: A SUCCESSION OF WIND-BLOWN DEPOSITS AND OCCUPATION FROM 2400 B.C. UP TO HISTORIC TIMES

In the earlier part of the Holocene a narrow coastal barrier was pushed landward by the encroaching sea, but around 3000 B.C. this movement came to an end and from then on new coastal barriers were formed on the sea side of the older ones, separated from these by strand flats. Rather low dunes (the Older Dunes, not higher than 10 m (33 ft.)) were formed on the barriers; in the fossil strand peat formation began when the groundwater reached the surface. The coastal aggradation was a rapid process. A belt of barriers of *c.* 6 km (4 miles) width was formed during the Subboreal and the process continued probably during the early Subatlantic. In Roman times the coastline lay some kilometres farther to the west than at present. The causes of the change in coastal processes at about 3000 B.C. are still rather obscure. The decrease in the rise of the sea-level is certainly one factor in this, but another must be the increase in the amount of sand available at the coast, either derived from the North Sea bottom (changed current patterns) or from the rivers, or both. The same factors must have played a role in the coastal degradation after the Roman period, which resulted in a smoothing of the coastline and the formation of the Younger Dunes.

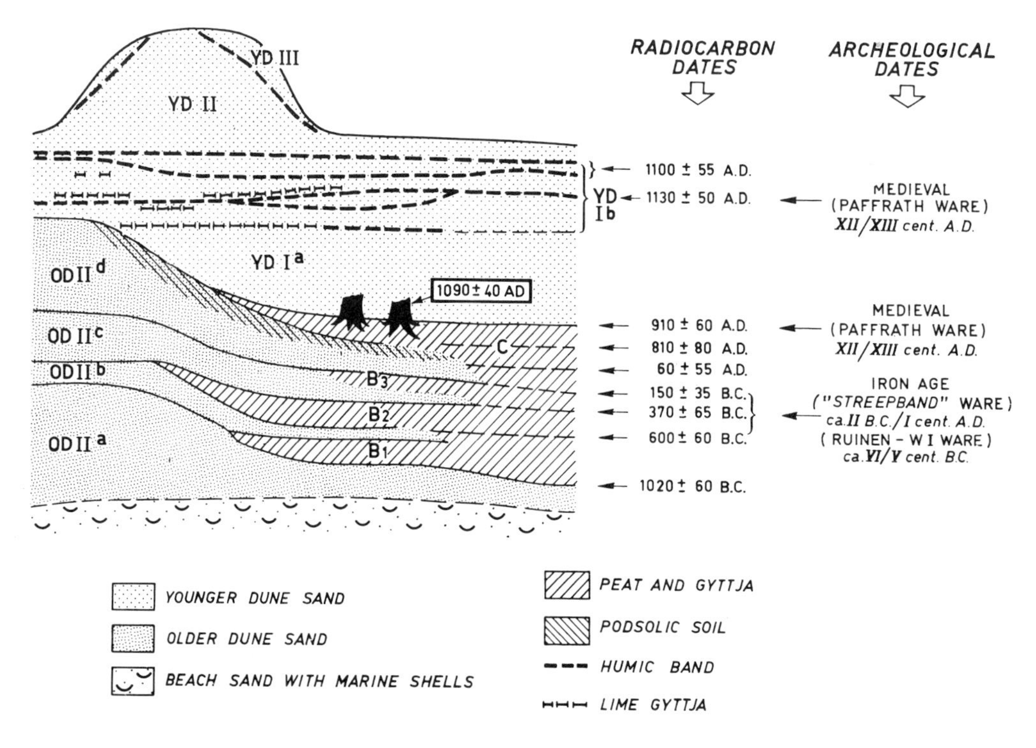


Fig. 53. Schematic representation of the succession of wind-blown sands, peat and soil formations in the dune area south-west of Haarlem. After S. Jelgersma and J. F. van Regteren Altena, *Geologie en Mijnbouw,* **48** (1969), 339.

The coastal barrier belt was broken into sections by some wide inlets at the mouths of the main rivers, where curved spits were formed. The coastline itself must have been rather open with a low halophytic vegetation, but the rows of Older Dunes further inland were covered with a deciduous forest, in which the oak dominated. On the peat-covered strand flats alder brushwood was found.

The dunes formed an attractive landscape for settlement. The main rivers and the levee deposits along their banks formed the routes through the immense peat bogs, by which they were accessible from the sandy hinterland.

The conditions for archaeological research are widely different within the district. Much has gone, since the dense present-day occupation is concentrated on the same dunes; the bulb fields were made (i.e. levelled and dug) on these sands before the time of archaeological interest. Most organic material decayed, because of the position of most culture layers well above the ground water table. But prehistoric sites can, on the other hand, have been beautifully preserved, when covered with wind-blown sand, and at low points conditions for preservation can be favourable, as in the Velsen area, excavated by J. F. van Regteren Altena. These sites enable us to sketch the following picture.

Traces of coastal occupation from the period before 3000 B.C. are irretrievably lost. Only one isolated axe might belong to the period 3000–2400 B.C. Our picture of the occupation starts with some Vlaardingen settlements on the dunes of the oldest barrier. These were already covered with forest, and the actual coastline lay 2 km (1.3 miles) farther west. This seems to be characteristic for all coastal occupation: there are no settlements known from the actual coastlines. Nor are they to be expected there, and if any were ever present there is little chance that their remains were preserved in that unstable environment. I will not go into detail about the subsistence economy in these settlements. At all sites cattle raising was most important in all periods, with pigs and sheep in widely varying ratios in the second and third positions. Hunting, especially of roe deer and red deer, was of minor importance in the Neolithic and early Bronze Age, and of no importance at all in the Iron Age and later.

The geological history is documented in the stratigraphy of the Older Dunes. A sequence of wind-blown sands has been observed, separated by soil profiles, peat or gyttja layers, that represent dormant phases, well-dated by a great number of C14 dates and by domestic refuse of prehistoric settlements.

The three oldest 'A' levels are only present in the older, eastern parts of the dunes and correlated with Vlaardingen—All Over Cord Beaker, Late Bell Beaker–Early Bronze Age, and Middle Bronze Age material respectively. The two lower 'B' levels contain Late Bronze Age sherds (1000–700 B.C.) and Middle Iron Age pottery (Ruinen/Wommels I–III ware, 600–450 B.C.). The upper B-level was formed before Roman times.

From this it seems preferable to speak of continuous occupation in spite of short breaks. The periods of sand displacement, although dominating the sections, were relatively brief. It seems best to infer continuous occupation, interrupted by intervals of one or two centuries of general dune formation. The sequence is in fact one of periods with a high ground water table, a vegetational cover and human occupation, alternating with periods with a lower ground water level, destruction of vegetation and a lack of archaeological finds, which in view of the environmental situation, is of no great surprise. There seems to be a correlation with the transgression cycles in a general sense. Does this mean that the marine sequence is governed by a climatic sequence? One can only answer in the affirmative when one can prove that the groundwater changes in the dunes, reflecting variations in the precipitation–evaporation ratio, are caused by natural processes. But prehistoric and early historic man seems to have played an important role, by his deforestation, grazing of cattle and other agricultural activities. In a dune landscape, covered by a natural vegetation, minor variations in precipitation would not have been disastrous. But the dunes are a vulnerable district and, when the vegetation is disturbed, dune formation is easily started. In this way man may have had a considerable influence on the occurrence of phases of dune formation, and especially in later prehistory and early history when more extensive areas were occupied and cultivated. I estimate human influence on this process before the Iron Age to be very low.

With regard to the transgression/regression cycles it is suggested that the coastal inlets might have been (partially) blocked by wind-blown sand during a phase of dune formation and that this might be a cause of the start of a regression phase, which is, in fact, primarily a restriction of marine ingressions into the intracoastal area. This explanation cannot be applied, however, to the occurrence of transgression/regression cycles along other coasts.

Summarising all data and considerations, I think we can state that the cyclicity in the dunes is primarily the result of variations in precipitation, while in later prehistory and early history the effect will have been enlarged as a result of human destruction of the natural vegetation. The same climatic changes are responsible for the transgression/regression cycles, with a higher storm and stormflood frequency and a higher precipitation in the transgression phases.

References: Glasbergen and Addink-Samplonius 1965; Glasbergen *et al.* 1967–8; Jelgersma and van Regteren Altena 1969; Jelgersma *et al.* 1970; Modderman 1960–1; van Straaten 1965; Zagwijn 1965.

### 5. WESTFRISIA — BRONZE AGE OCCUPATION ON SALT MARSH DEPOSITS 1300–700 B.C.

The district of Westfrisia is built up of salt marsh and creek deposits, that were formed during the transgressive phases Calais IV and Dunkirk O in a system of tidal creeks, all connected with a main sea-arm, which had penetrated far inland behind the inlet in the belt of coastal barriers near Egmond. These sediments were deposited up to a relatively very high level, and can be ascribed to an elevation of high waters in this system, especially when the minor creeks were blocked at the beginning of the sedimentation phase. Bronze Age remains (barrows) were already known in 1942, and later settlements from the same age were discovered during the soil survey by Ente (1963). New discoveries since that time demonstrated the richness of the region in this respect, but this has been threatened in recent years by a large scale reallotment. Extensive fieldwork and excavations were carried out by the Institute for Pre- and Protohistory, Amsterdam, and by the State Service for Archaeological Investigations. Amersfoort: a survey ('Landssaufnahme') of 3600 ha (8700 acres) was made, an 18 ha (45 acres) excavation completed near Bovenkarspel and a monument of 70 ha (175 acres)

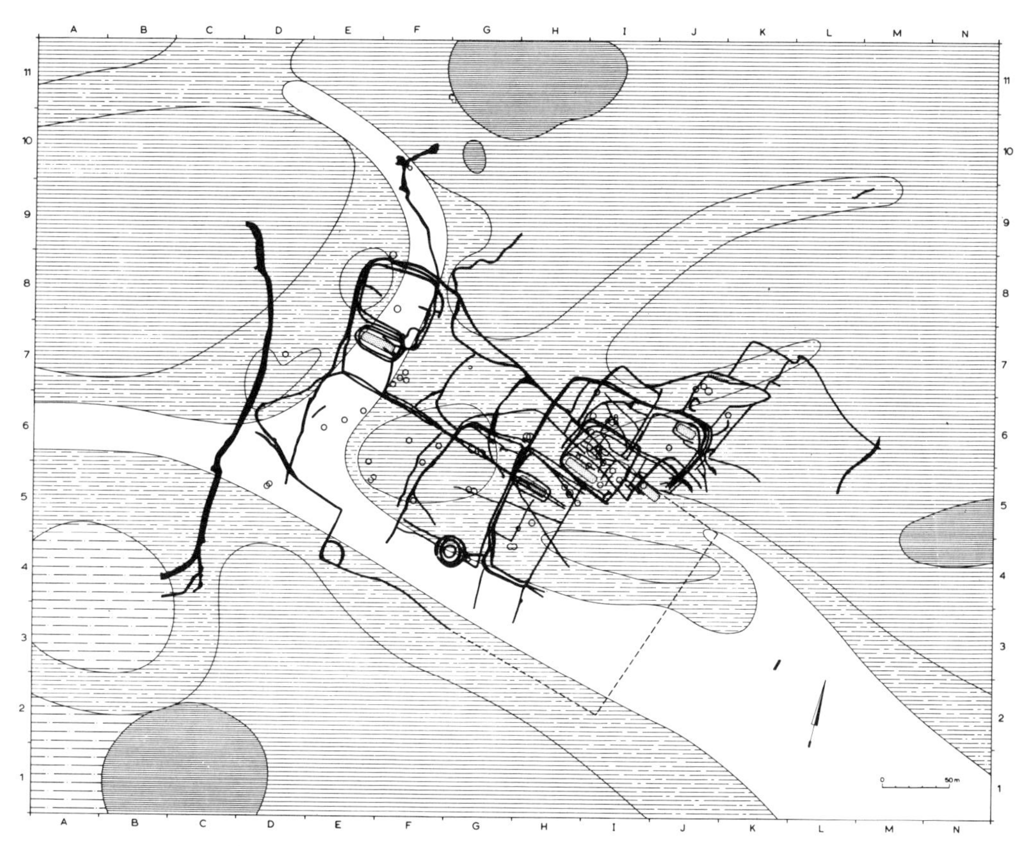


Fig. 54. Hoogkarspel-Watertower, Westfrisia. Idealised excavation plan, projected on the soil map, showing numerous successive ditch systems in close relationship to the soil conditions. The settlement, dated *c.* 1000 B.C., comprised probably one or two farmyards, rebuilt from time to time during the occupation, which itself lasted a few centuries. After J. A. Bakker *et al.* 1977.

- sand
- sandy clay
- sandy clay or clay
- clay or sandy clay
- clay
- housesite
- main ditches and drains
- granary drains
- largest possible extent of one holding

saved from destruction. A fairly complete picture of Bronze Age Westfrisia has been obtained.

There are a few beaker settlements on deposits of Calais IV[a]—age to the west of the district proper: Aartswoud (zigzag beaker, *c.* 2100 B.C.) and Oostwoud (Early Bell Beaker, *c.* 1900 B.C.). Bronze Age occupation concentrated more to the east, in the higher part of the region, and started not before 1300 B.C. The inlet at Egmond was probably narrowed or even closed at that time, and drainage conditions in eastern Westfrisia improved. Moreover, a change to freshwater conditions took place and an initial compaction, resulting in a landscape with a very low (less than 1 m (3 ft.) relief. The fields, measuring ¼–1 ha (½–2½ acres) were laid out on the sandy creek ridges (the main ones were more than 100 m (330 ft.) wide) and were surrounded by ditch systems which were frequently cleaned and altered. They were not very deep and had a drainage function only in times of a high water table. These fields were ploughed with the *ard*, as is revealed by extensive plough marks. The farms were built next to the fields, high on the flanks of the ridges. Some dozens of house plans are known, all of the three-aisled type, measuring, on average, 25 m (80 ft.), with a maximum width of 6 m (20 ft.). In one half, cattle stalls are presumed by analogy with similar houses at Elp, but they were not preserved. Most houses were surrounded by a ditch, to give better drainage to the yard in the wet season. In the lower terrain, vegetation had changed only slightly: treeless natural pasture lands were available for grazing, with lakes at the lowest places. Numerous small circular ditches and rings of pits mark places where most probably fodder, such as marsh-hay or straw, was stored, or perhaps even the harvest. No traces of wooden granaries were found. The expensive timber for the houses must have been brought from a great distance.

Permanent settlement and a fully agrarian, self-supporting economy seem evident. Cattle, a small breed, are represented by *c.* 75% in the refuse and were most important, with sheep second and pigs third. Grain (Emmer wheat and Six-row barley) and flax were grown. Some shellfish were collected at some distance in the tidal area. An occasional wild animal (elk, red deer, beaver, otter and others) was killed; these might have accidentally wandered into the area, or have been hunted elsewhere.

A Middle Bronze Age occupation phase, dated between 1300 and 1000 B.C. and characterised by simple, bucket-shaped pottery, can be separated from a Late Bronze Age phase between 850 and 700 B.C. No cause for this occupation hiatus has yet been found. In this second phase groundwater conditions worsened. People retreated to the highest points of the area and raised the yards with soil, dug from systems of surrounding ditches, which at the same time improved drainage. But in due course, as the sea-level and also the water table continued to rise, the land proved to be too low and, especially, too wet for cultivation and the raising of a yearly crop.

References: Bakker *et al.* 1959–68; Bakker *et al.* 1977; Buurman and Pals 1974; Clason 1967; Ente 1963; van Giffen 1944; van Giffen 1961; Modderman 1964; Modderman 1974; van Regteren Altena 1976; van Regteren Altena *et al.* 1977.

## 6. THE 'TERPEN' DISTRICT OF GRONINGEN AND FRIESLAND — IRON AGE AND LATER OCCUPATION OF SALT MARSHES, 500 B.C.–A.D. 1000

As a result of the progressive decrease in the rise of sea-level the salt marshes of the end of the Dunkirk I transgressive phase were the first that ceased to be marshy shortly after their formation. No peat cover was formed. This must be the main reason for the occupation of the northern coastal district starting at this time, i.e. about 500 B.C. In the first half of the Iron Age settlements were founded on the higher, sandier, and better drained ridges (the marsh bars) and creek (curs.) levees, in Groningen as well as Friesland. It has been asked where the colonists came from, and why they left their regions of origin. Waterbolk demonstrated that in this phase the conditions for agriculture on the Drenthe plateau deteriorated; the Celtic field systems suffered from wind erosion, although this now appears not to have been in such a catastrophic sense as was originally suggested. The pottery of the oldest marsh settlements is similar to that of the settlements in the sand district: the Ruinen-Wommels I/II ware of the Zeijen culture. The region of origin of some others might have been Westfrisia, which became uninhabitable just prior to this period.

The marshes lay in a relatively protected situation, behind extensive tidal flats, which very probably were separated from the open sea by a row of barrier islands, very similar to the present geography. Palaeobotanic research has revealed that the vegetation was entirely halophytic, with *Salicornia maritima* and *Juncus gerardii* as the dominant plants, which indicates regular, but non-destructive, floodings. The rushes offered rich, natural pasture lands: cattle breeding was the main activity, to provide milk, meat, hides and tractive power. Sheep were second in importance in their livestock, and only a few pigs were kept. This sheep/pig ratio is (as for earlier situations in Westfrisia) characteristic for the treeless landscape. Some hunting took place of aurochs, elk, bear, red deer and other animals, either on the wooded sands to the south, or whenever they wandered into the marsh district. Seals were caught along the coast. Crop cultivation was rather hazardous: barley, *Linum* (flax) and other crops, such as *Brassica campestris, Camelina sativa* and *Daucus carota* (carrot) could withstand the floodings, but no wheat and pulses could be grown.

The occupation certainly was permanent and the people self-supporting. Serious difficulties had to be overcome: apart from the adaptation of the agricultural system, there was the lack of water, fuel and timber. Rainwater was collected in small ponds, on top of artificial mounds (see below), and timber was found on the sands to the south where fuel could also be collected; otherwise dried dung could serve as fuel. The major limiting factor on occupation was not in the settlement itself, but in the possibilities of raising an annual crop and grazing cattle, on land not too frequently flooded with salt water.

Shortly after the first colonisation a new transgressive phase started (Dunkirk $I^{b.1}$), which, although of modest extent and not destructive, had its effect on the living conditions. Some settlements were left (Middlestum, Hatzum, Jemgum), at other sites people demonstrated their remarkable adaptability by the construction of the first *terpen*, artificial mounds of sods, on which the farms were built (Ezinge).

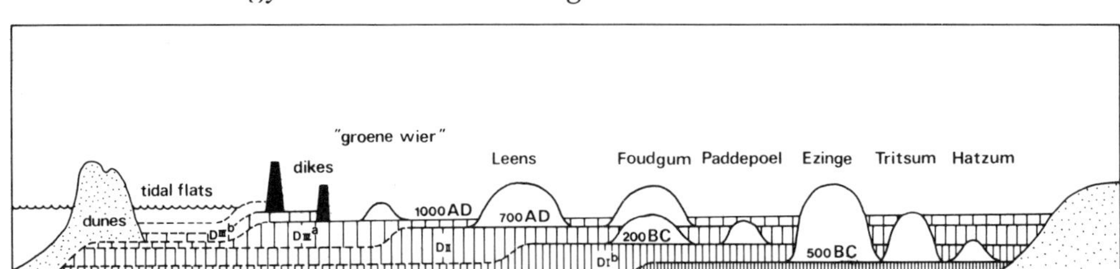


Fig. 55. Schematic north–south section through the Groningen *terpen* and salt marsh district, showing the successive marsh deposits, the shift of newly founded 'terpen' northward and the extension or abandonment of the older ones. Redrawn after H. T. Waterbolk, in J. W. Boersma ed. 1970.

The formation of salt marshes continued, linked to the transgression/regression cycles. Four major phases can be distinguished, each followed by new colonisation of the slightly higher, fresh pasture lands on the seaward side of the old ones, and mark the beginning of a new generation of *terpen*. The older ones were raised and extended and sometimes fused, to make room for more houses and for crop cultivation on the slopes. But at the same time the oldest and lowest marshes, far inland, sometimes became too marshy. Settlements in the Roman period are known to have been abandoned for this reason at the end of the third century and subsequently covered by sediments (Paddepoel). The following general sequence of events in the northern coastal district can now be made out:

(a) Dunkirk $I^a$ salt marsh formation, followed by surface settlements on marsh bars and levees, dated by Ruinen/Wommels I/II pottery, 500 B.C.
(b) Dunkirk $I^{b1}$ transgressive sub-phase. Settlements are left or small initial *terpen* are constructed. Ruinen/Wommels III pottery of the 'protofrisian culture', 400–200 B.C. First generation of *terpen*.
(c) A short regressive interval *c.* 200 B.C. with extension of occupation and start of second generation of *terpen* during the following Dunkirk $I^{b2}$ transgressive subphase. '*Streepband* pottery' of the Frisian Culture, with its continuation into Roman times.
(d) Friesland and Groningen were not occupied by the Romans, but there existed trade relationships with the Roman provinces south of the Rhine. In the following centuries the sequence of events is rather obscure. Occupation continued during the Dunkirk II transgressive phase, but in view of the modest number of finds there may have been a decrease in population.
(e) A third generation of *terpen* was founded on the Dunkirk II marshes after *c.* A.D. 700.
(f) Shortly before the embankments started some small *terpen* were raised on Dunkirk $III^a$ marshes: a fourth generation. But the embankment of the salt marsh area made further raising and extension of the *terpen* unnecessary. Their growth stopped and farms were founded on the surface of the flat country, near their fields.

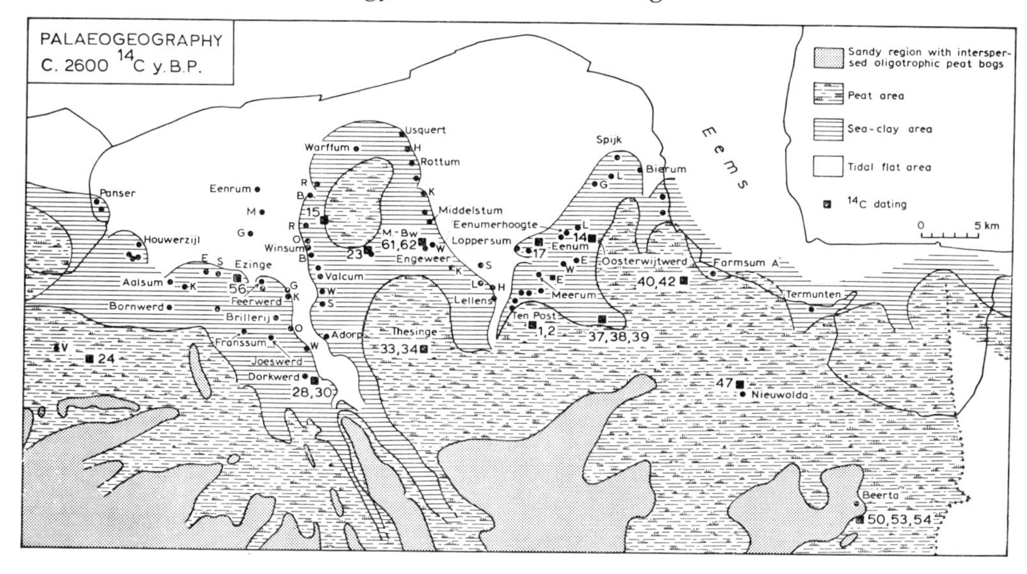


Fig. 56. Palaeogeographic map of the Groningen coastal area shortly after the Dunkirk I[a] phase of salt marsh formation and the first colonisation of the fresh deposits by the farmers of the 'protofrisian' culture, who shortly afterwards must have built the first *terpen*. After W. Roeleveld 1974 (fig. 63).

References: Boeles 1951; Boersma 1970; van Es 1968; van Giffen 1936; van Giffen 1940; Roeleveld 1974; Waterbolk 1965–6; van Zeist 1974.

## 7. THE OCCUPATION SEQUENCE IN THE WESTERN NETHERLANDS AFTER THE ROMAN PERIOD

In the Roman period a widespread occupation took place over the whole of the western Netherlands (the peat districts excepted), and also on the barriers. At the end of the third century there was a sharp decline in number of sites, which certainly reflects a decline in population, but the lack of datable (imported) material may be partly responsible for this impression. From the fifth and sixth centuries no finds at all are known and the population must have been very thin, or even absent, but from the end of the sixth century onward the number of finds and sites increases.

The oldest are situated at the mouths of the rivers Scheldt (Domburg), Meuse (Naaldwijk, Monster *inter alia*) and Rhine (Rijnsburg, *inter alia*). The main reason for this occupation hiatus was formerly found in the transgression phase Dunkirk II, but it appears that many of the deposits dated to this phase are in fact later (Dunkirk III[a]), since twelfth-century house sites were found below these clays at several points. So political reasons and an economic collapse after the Roman departure are now considered to be a better explanation and a major cause.

From the Carolingian period onwards a steady increase in the number of sites

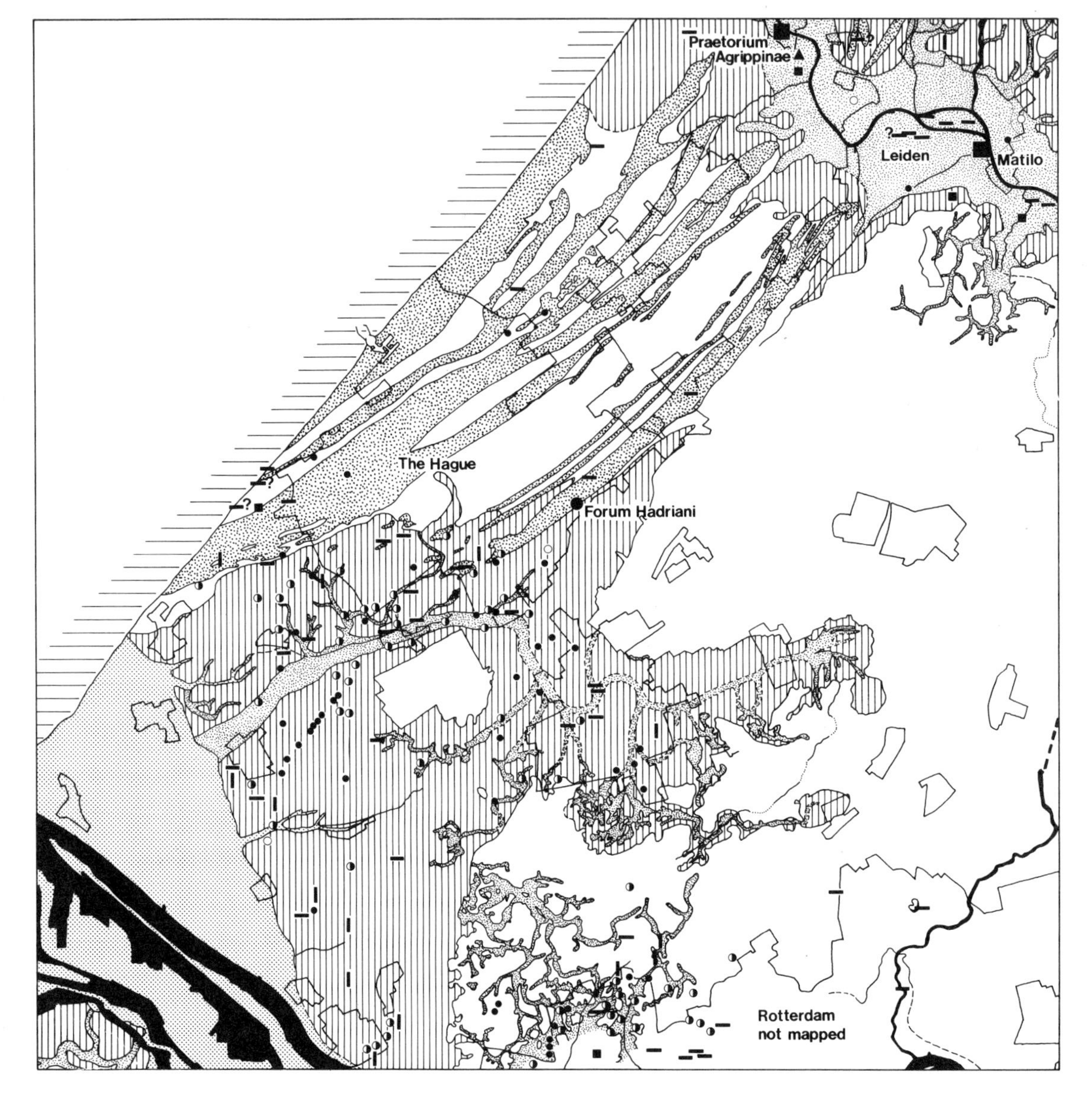


Fig. 57. Geological map of the coastal area between Leiden and Rotterdam and its Roman occupation. Scale 1:200.000.

Geology

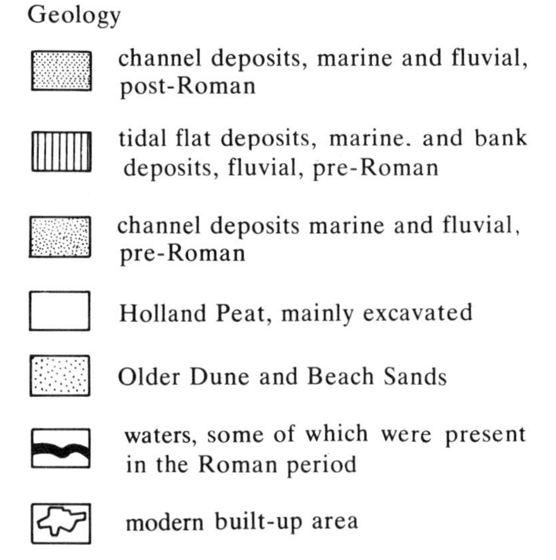

- channel deposits, marine and fluvial, post-Roman
- tidal flat deposits, marine. and bank deposits, fluvial, pre-Roman
- channel deposits marine and fluvial, pre-Roman
- Holland Peat, mainly excavated
- Older Dune and Beach Sands
- waters, some of which were present in the Roman period
- modern built-up area

Archaeology

- ■ Roman fort
- ● Civitas capital
- ▪ settlement, mainly finds of Roman material
- ◐ settlement, finds of Roman and native-Roman material
- ○ settlement, mainly finds of native-Roman material
- ▬ site, *c.* five finds or less, Roman material
- • id. Roman and native-Roman material
- ▮ id. native-Roman material
- ▲ Roman cemetery
- ? location uncertain

reflects a rapidly growing population. The colonisation and reclamation of areas, unoccupied for centuries or even millennia, as in the case of the peat districts, was stimulated for political reasons by the Count of Holland and the Bishop of Utrecht and in the thirteenth century practically the whole of Holland was taken into (agricultural) use, embanked and artificially drained in a modest way. From this time onwards a sequence developed of improved drainage, raised embankments, and an increased danger of flooding. The natural equilibrium was broken. Transgression phases are obscured by other phenomena, such as the quality of the maintenance of the dykes, the sedimentation along the rivers between the dykes and the raising of floods between the embanked sea arms.

In the south-west, the province of Zeeland, the greater part of the peat district was destroyed by the sea between the Roman period and the eleventh century. Only minor parts of the 'old landscape' are preserved in the central parts of some of the islands, which came into existence by the embankment of fresh salt marsh deposits. A similar, and even more complete, destruction took place in the same period in the northern part of Holland. It may not be accidental that in both districts the belt of coastal barriers curved farthest seaward, while we know from the Holland Older Dunes landscape that in this period coastal degradation took place, resulting in a straightening of the coastline.

In the dune landscape occupation started again after the hiatus of the fifth and sixth centuries. This is confirmed by pollen diagrams, which show a complete vegetation recovery of the dunes, followed by fresh reclamations gradually leading to a disappearance of the forest vegetation. This might be one of the causes of the formation of the Younger Dunes.

The main cause must, however, be a change in coastal processes (a change of the currents?), resulting in the degradation mentioned above and the straightening of the coast, so making available the huge quantities of sand, from which the Younger Dunes were formed. These changes must have started sometime between the fourth and twelfth century A.D., since sherds dating from the twelfth century were found in a fossil arable land over which the sands of the oldest phase of the Younger Dunes was blown. Where these dunes were blown landward, habitation and agriculture became impossible, and people had to retreat landward.

References: Besteman 1974a; Besteman 1974b; Bloemers 1978; Jelgersma *et al.* 1970; Sarfatij 1971; Trimpe Burger 1973.

*I am indebted to Miss Linda Whitaker for struggling with the original English text. MS. closed May 1978.

## BIBLIOGRAPHY

Bakker, J. A. *et al*. 1959–68. Opgrayingen te Hoogkarspel I–V *West-Frieslands Oud en Nieuw,* **26** (1959), 158–91; **33** (1966), 168–75, 176–224; **34** (1967), 202–28; **35** (1968), 192–9.

—— 1977. Hoogkarspel-Watertoren, towards a reconstruction of ecology and archaeology of an agrarian settlement of 1000 B.C., in *Ex Horreo = Cingula,* **IV,** Amsterdam, 187–225.

Bennema, J. 1954. Bodem- en zeespiegelbewegingen in het Nederlandse kustgebied, thesis Wageningen; also: *Boor en Spade,* **7,** 1–96.

Besteman, J. C. 1974a. Carolingian Medemblik. *Ber. R.O.B.,* **24,** 43–106.

—— 1974b. Frisian salt and the problem of salt-making in North Holland in the Carolingian period. *Ber. R.O.B.,* **24,** 171–4.

Bloemers, J. H. F. 1978. Rijswijk (Z.H.), 'De Bult'. Eine Siedlung der Cananefaten. *Nederlandse Oudheden,* **8.**

Boeles, P. C. J. A. 1951. *Friesland tot de elfde eeuw,* 's-Gravenhage.

Boersma, J. W. (ed.). 1970. *Terpen, mens en milieu,* Haren (Gr.).

Buurman, J. and Pals, J. P. 1974. Some remarks on prehistoric flax in the Netherlands. *Ber. R.O.B.,* **24,** 107–11.

Clason, A. T. 1967. Animal and man in Holland's past, thesis Groningen; also: *Palaeohistoria,* **13.**

Ente, P. J. 1963. Een bodemkartering van het tuinbouwcentrum 'De Streek', thesis Wageningen; also *Versl. Landb. Onderz.* Nr. **68,** 16.

Glasbergen W. and Addink-Samplonius, M. 1965. Laat-Neolithicum en Brontijd te Monster (Z.H.). *Helinium,* **5,** 97–117.

—— *et al.* 1967–8. Settlements of the Vlaardingen culture at Voorschoten and Leidschendam, I–III. *Helinium,* **7,** 3–31, 97–120; **8,** 105–30.

Hageman, B. P. 1969. Development of the western part of the Netherlands during the Holocene. *Geologie en Mijnbouw,* **48,** 373–88.

Jelgersma, S. 1961. Holocene sea level changes in the Netherlands, thesis Leiden; also: *Meded. Geol. Sticht.* C VI, 7.

—— 1966. Sea-level changes during the last 10,000 years, in Royal Meteorological Society *Proceedings of the International Symposium on World Climate from 8000 to 0* B.C., 54–71.

—— and van Regteren Altena, J. F., 1969. An outline of the geological history of the coastal dunes of the western Netherlands. *Geologie en Mijnbouw,* **48,** 335–42.

—— *et al.* 1970. The coastal dunes of the western Netherlands; geology, vegetational history and archaeology. *Meded. Rijks Geol. Dienst* NS No. **21,** 93–167.

Jong, J. D. de 1967. The Quaternary of the Netherlands, in K. Rankama (ed.), *The Geological Systems – The Quaternary,* Vol. 2, New York, 1967, 302–426.

—— 1971. The scenery of the Netherlands against the background of Holocene geology; a review of recent literature. *Revue de géographie physique et de géologie dynamique,* **13,** 143–62.

Louwe Kooijmans, L. P. 1970–71. Mesolithic bone and antler implements from the North Sea and from the Netherlands. *Ber. R.O.B.,* **20/21,** 27–73.

—— 1974. The Rhine/Meuse Delta: four studies on its prehistoric occupation and Holocene geology, thesis Leiden; also: *Oudheidk. Meded. Leiden,* **53/59** (1972/3), and *Analecta Praehistorica Leidensia,* **7** (1974).

—— 1976a. The Neolithic at the Lower Rhine, its structure in chronological and geographical respect, in: Acculturation and continuity in Atlantic Europe. *Diss. Arch. Gand.* **16,** 150–73.

—— 1976b. Local developments in a borderland, a survey of the Neolithic at the Lower Rhine. *Oudheidk. Meded.,* **57,** 227–97.

—— 1976c. Prähistorische Besiedlung im Rhein-Maas-Deltagebiet und die Bestimmung ehemaliger Wasserhöhen. *Probleme der Küstenforschung im südlichen Nordseegebiet,* **11,** 119–43.

Modderman, P. J. R. 1960–1. De Spanjaardsberg; voor- en vroeghistorische boerenbedrijven te Santpoort. *Ber. R.O.B.,* **10/11,** 210–62.

—— 1964. Middle Bronze Age graves and settlement traces at Zwaagdijk. *Ber. R.O.B.,* **14,** 27–36.

—— 1974. Een drieperiodenheuvel uit de Midden Bronstijd op het Bullenland te Hoogkarspel. *West-Frieslands Oud en Nieuw,* **41,** 251–59.

Newell, R. R. 1973. The post-glacial adaptations of the indigenous population of the northwest European plain, in: *The Mesolithic in Europe,* 399–440. Warszawa.

Pons, L. J. *et al.,* 1963. Evolution of the Netherlands coastal area during the Holocene. *Verhand. Kon. Ned. Geol. & Mijnbouwk. Genootsch.,* 197–207.

Roeleveld, W. 1974. The Holocene evolution of the Groningen marine clay district, thesis Free University of Amsterdam; also: *Ber. R.O.B.,* **24,** supp.

Sarfatij, H. 1971. Friezen, Romeinen, Cananefaten. *Holland,* **3,** 33–47, 89–105, 153–79, also: *R.O.B.* reprint nr. 33.

Swifterbant Contributions 1–8, by various authors, in *Helinium,* **16/17,** 1976–77.

Trimpe Burger, J. A. 1973. The islands of Zeeland and South Holland in Roman times. *Ber. R.O.B.*, **23,** 135–48.

Van Es, W. A. 1968. Paddepoel, excavations of frustrated terps, 200 B.C. – A.D. 250. *Palaeohistoria,* **14,** 187–352.

—— 1974. Archaeology in the Netherlands. *Rescue News,* **8,** 8–10.

Van Giffen, A. E. 1936. Der Warf in Ezinge, Provinz Groningen und seine westgermanischen Häuser. *Germania,* **20,** 40–47.

—— 1940. Die Wurtenforschung in Holland. *Probleme der Küstenforschung im südlichen Nordseegebiet,* **1,** 70–86.

—— 1944. Grafheuvels te Zwaagdijk. Corrected and augmented reprint from *West-Frieslands Oud en Nieuw,* **17,** 121–231.

—— 1961. Settlement traces of the Early Bell Beaker culture at Oostwould (N.H.). *Helinium,* **1,** 223–8.

Van der Plassche, O. forthcoming. Sea-level movements in the mid-western Netherlands, thesis Free University of Amsterdam.

Van Regteren Altena, J. F. 1976. Polder Grootslag en Bovenkarspel, Archeologische Kroniek van Noord-Holland over 1976 (ed. P. J. Woltering). *R.O.B.* reprint nr. 89, 187–92.

—— *et al.* 1962–3. The Vlaardingen culture. *Helinium,* **2,** 3–35, 97–103, 215–43; **3,** 39–54, 97–120.

—— 1977. Bronze Age clay animals from Grootebroek, in *Ex Horreo = Cingula,* **IV,** Amsterdam, 241–54.

Van Straaten, L.M.J.U. 1965. Coastal barrier deposits in South and North Holland, in particular in the areas around Scheveningen and IJmuiden. *Meded. Geol. Sticht.,* NS No. **17,** 41–75.

Van der Waals, J. D. 1972. Die durchlochten Rössener Keile und das frühe Neolithikum in Belgien und in die Niederlanden. *Fundamenta,* **A3,** VA, 153–84.

Van Zeist, W. 1968. Prehistoric and early historic food plants in the Netherlands. *Palaeohistoria,* **14,** 42–173.

—— 1974. Palaeobotanical studies of settlement sites in the coastal areas of the Netherlands. *Palaeohistoria,* **16,** 223–371.

Waterbolk, H. T. 1965–66. The occupation of Friesland in the prehistoric period. *Ber. R.O.B.,* **15/16,** 13–35.

Zagwijn, W. H. 1965. Pollen-analytic correlations in the coastal barrier deposits near The Hague (The Netherlands). *Meded. Geol. Sticht.,* NS No. **17,** 83–8.

# Post-glacial Environmental Change and Man in the Thames Estuary: a Synopsis

Dr R. J. Devoy

Situated at the southern end of the North Sea, the Thames estuary forms an eastward facing funnel-shaped area. Travelling westwards, the estuary narrows from ~27 km (16.7 miles) at its mouth to ~3 km (1.86 miles) near Cliffe Marshes (fig. 58). Its location and shape make it a focal point for concentrating the effects of southward moving storm surge phenomena. As a result this area often records maximal coastal water levels during the passage of such features. Storm surges along shorelines of the southern North Sea are caused by a combination of factors. Initially, they arise from the positioning of a deep low pressure centre over the North Sea, with its associated northerly winds backing up water to the south, and also from the occurrence of high spring tide levels. These conditions in conjunction with increased river discharges, produced by the rainfall often associated with the depression, result in the estuary forming a high risk zone for flooding:

> 'The day was the 18th November, 1897, and the wind switched suddenly into the opposite direction from that it had been blowing the day before. The day was overcast and dull, and the morning tide had ebbed so far out that no water could be seen in the creek. After dinner the tide suddenly appeared far down the creek and rushing up with a ridge of white foam at its front edge. Very soon it was breaking over the sea walls, overflowing low-lying roads, houses and buildings. The marshes of Great Barksoar Farm were flooded and many sheep were drowned in spite of great efforts of Mr Hanmer and his farm hands . . .'

This dramatic account by Mr A. Hawkins (Evans 1953) evidences a storm surge at Lower Halstow on the river Medway and its destruction of the man-made environment. In terms of the coastal evolution of eastern England this was not an exceptional occurrence. Records from East Anglia and the Thames frequently note similar catastrophes. Some observations from Essex and north Kent date from the beginning of the 11th century and evidence the increasing frequency of such phenomena from the 13th century onwards (Evans 1953; Grieve 1959). In

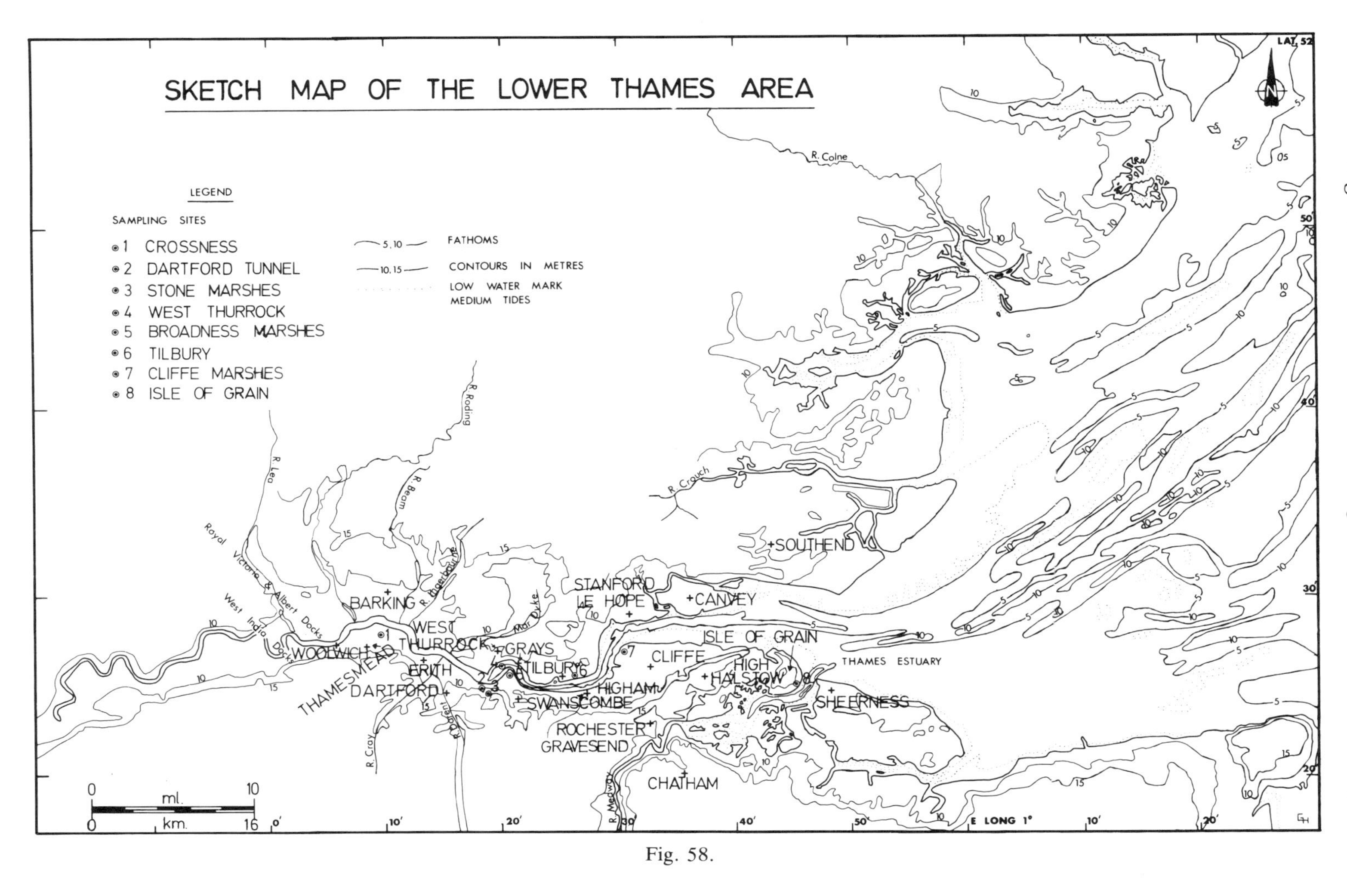
SKETCH MAP OF THE LOWER THAMES AREA
LEGEND
SAMPLING SITES
1 CROSSNESS
2 DARTFORD TUNNEL
3 STONE MARSHES
4 WEST THURROCK
5 BROADNESS MARSHES
6 TILBURY
7 CLIFFE MARSHES
8 ISLE OF GRAIN
5, 10 FATHOMS
10, 15 CONTOURS IN METRES
LOW WATER MARK MEDIUM TIDES
R. Colne
R. Roding
R. Lea
R. Beam
R. Crouch
SOUTHEND
Royal Victoria & Albert Docks
West India Docks
BARKING
STANFORD LE HOPE
CANVEY
WEST THURROCK
GRAYS
ISLE OF GRAIN
THAMES ESTUARY
WOOLWICH
THAMESMEAD
ERITH
TILBURY
CLIFFE
HIGH HALSTOW
DARTFORD
HIGHAM
SWANSCOMBE
SHEERNESS
ROCHESTER
GRAVESEND
R. Cray
R. Darent
R. Medway
CHATHAM
0 ml. 10
0 km. 16
E LONG 1°
LAT 52
GH

Fig. 58.

the 1897 event high tide in the area rose to ~+4.6 m O.D. (15 ft.) More recently in the 1953 storm surge, which caused widespread flooding and damage to urban infrastructure, and also in the less disastrous storm of 1978, water levels exceeded ~+5.2 m O.D. (17 ft.) at London Bridge. The height of all these surges easily overtops the general height of areas bodering the Thames estuary. Most of these lie below the height of mean high water mark of spring tides (M.H.W.S.T.), which varies between +2.8 m O.D. (9 ft.) at Sheerness to +3.9 m (12.8 ft.) at London Bridge (Admiralty Tide Tables 1977). Serious flooding is only currently prevented by the presence of embankments, although the construction of an integrated flood alleviation network is under way (Horner 1972). The trend in height of surge phenomena and of tidal levels overall has been one of a linear increase, since records began in this area in the nineteenth century (Rossiter 1972). The cause of this feature lies in part with the progressive definition of estuary shape since 1830, a long term relative eastward tectonic subsidence of south-east England, and a recent increase in the frequency of easterly gales, coupled with the continuing rise of post-glacial ocean volume (Rossiter 1972; Devoy 1979).

The pattern of these inundations, produced as annual or longer return period surges, rarely leaves any permanent sedimentary record, although they are of vital concern to man with high capital investment in this low-lying area. On a longer time scale, marine incursions of greater magnitude and duration have been recorded in the stratigraphy, leaving clay, silt and sand sequences. In the Thames estuary deposits spanning the last 10,000 years (Flandrian stage) of the Quaternary show the occurrence of relative sea-level and coastal movements in alternating levels of biogenic and inorganic sediments. The biogenic deposits are composed of peats and gyttjas, representing the removal of the marine influence with the growth of fresh and brackish water plant communities. Five regression phases (Tilbury I–V), with the possibility of a sixth, have been recognised in the inner estuary, based upon Tilbury as the type site (Table 1). The inorganic levels are formed by blue/grey clays and silts <60 μm in diameter, representing marine/brackish water conditions. From a biostratigraphic study and C14 dating of the contact points between these alternating layers, five main marine transgressions have been recognised (Devoy 1977) (fig. 59). These show a relative sea-level rise for M.H.W.S.T. from −25.5 m O.D. (−83.7 ft.) at ~8200 years bp (Isle of Grain) to +0.4 m O.D. (*c.* 1.3 ft.) at ~1750 years bp (Tilbury) (see Table 2).

At Canvey Island, Cliffe, Cooling and St Mary's Marshes and in the river Medway area (fig. 58), interleaved organic and inorganic sediments have been

Table 1. Variation in depth of the biogenic deposits in the Thames estuary

| *Type site Code* | *Height taken to top of each level* | |
|---|---|---|
| T V | +0.40 to − 0.90 m O.D. | (1.31 to −2.95 ft.) |
| T IV | −0.80 to − 1.80 m O.D. | (−2.62 to −5.9 ft.) |
| T III | −1.90 to − 5.20 m O.D. | (−6.23 to −17.06 ft.) |
| T II | −6.80 to − 10.07 m O.D. | (−22.3 to −33 ft.) |
| T I | −12.23 to − 25.53 m O.D. | (−43.39 to −83.74 ft.) (I. of G.) |

Table 2. Flandrian marine transgression sequences in the Lower Thames estuary

| | |
|---|---|
| Thames V | ~1750 years bp |
| Thames IV | 2600– / years bp |
| Thames III | 3850–2800 years bp |
| Thames II | 6575–5410 years bp |
| Thames I | 8200–6970 years bp |

recorded at similar levels (Evans 1953; Lake, Ellison, Henson and Conway 1975). The biogenic deposits here are generally much thinner, between 10 and 30 cm (3.93–11.8 in.) thick. These are composed predominantly of monocotyledonous and herbaceous plant remains with a high inorganic fraction present. Where these have been C14 dated, the height and radiometric data support the five main transgression and regression phases identified. However, the lack of detailed biostratigraphic evidence for these sites makes a close correlation of the Flandrian sequences difficult. Throughout the inner estuary these deposits overlie a late Devensian sand/gravel surface, in which the rivers Thames and Medway developed the final 'second' buried channel phase. This reaches a depth of −30 m O.D. (−98.5 ft.) in the Isle of Grain area, rising to −13 m O.D. (−42.6 ft.) at Tilbury and −5 m O.D. (−16.4 ft.) at Charing Cross (Lake *et al*. 1975). Subsidiary feeder channels along the course of the Thames leave residual sub-horizontal intefluve surfaces.

In the Foulness and Dengie Flats area of the outer estuary six transgression and five main regression phases have been recorded (Greensmith and Tucker 1971a.b, 1973). The late Devensian surface here similarly shows the formation of deep channels, together with the development of the now buried channel phases of the rivers Roach and Crouch to depths >−23 m O.D. (−75.4 ft.) Recognition of the subsequent marine transgressions is based upon evidence from a series of shell and sand ridges or 'cheniers', marsh retreat features and vertical changes in shell fauna. Evidence for regression again comes from a variety of sources, identification of overconsolidation layers, peat horizons and geosols. The first transgression, recorded at varying heights below −21 m O.D. (−68.8 ft.), is assigned to the period between 8900 and 7500 years bp. Transgression 2 ends by ~4000 years bp with a regression phase identified by over-consolidated layers at −11 to −7 m O.D. (−36 to −23 ft.) beneath Dengie Flats and Maplin Sands. A peat possibly formed close to high water mark at −1.5 m O.D. (−5 ft.) gives a date of 4959 ± 65 years bp for this regression and accords in time with Tilbury (T) III (fig. 59). Formation of shell banks at −7.5 m O.D. (−24.6 ft.) indicates the beginning of transgression 3. The lack of datable material makes the placing of C14 limits on subsequent phases difficult. Transgression 5 appears to begin at ~1400 years bp and correlates approximately with Thames (Th) V. Transgression 6 is placed at ~300 years bp. Partial disagreement on the timing, number and amplitude of the relative sea-level movements appears to occur between the inner and outer estuary. The cause for this may lie in differing environmental and sedimentological histories of the two areas, together with the variation in data sources used (Devoy, 1979).

Further north at Aldeburgh Marshes (Carr and Baker 1968) radiometric assays have been taken from peats formed close to the M.H.W.S.T. regime. These give

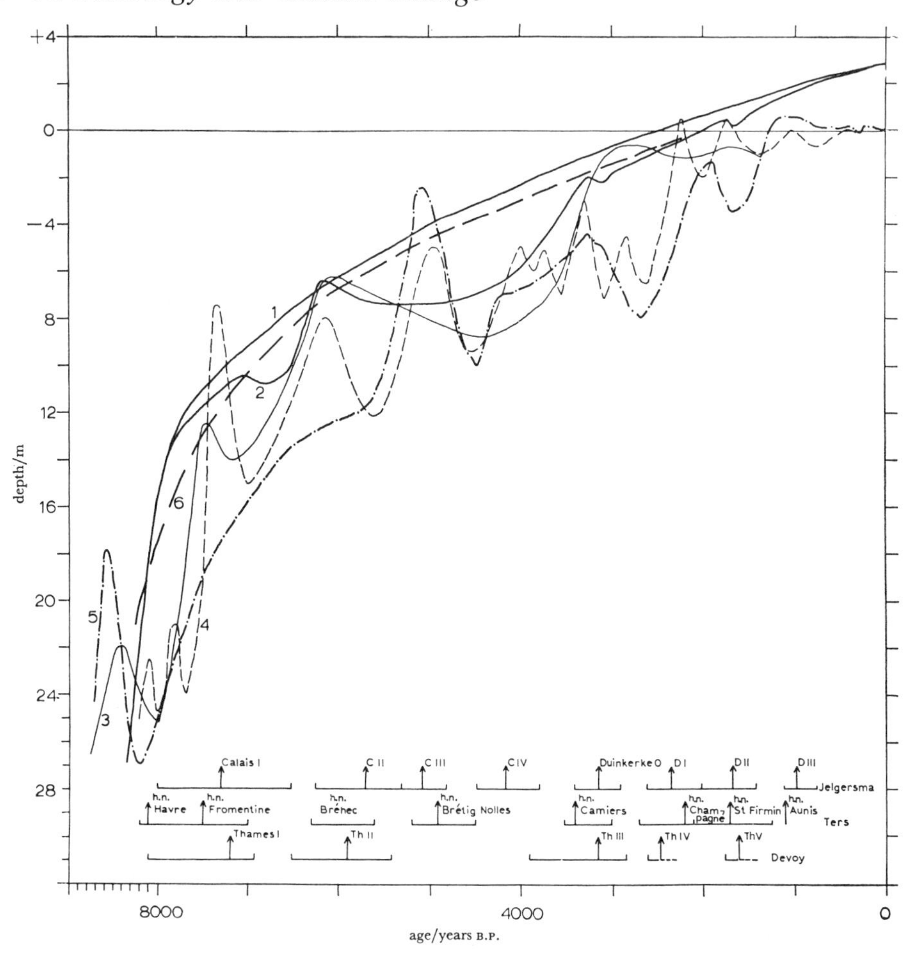


Fig. 59. Relative sea-level curves from the southern North Sea region and the Atlantic coast of France, derived from Flandrian sediments. 1. Thames mean curve (Devoy 1979); 2. Tilbury (Devoy 1979); 3. Morzadec-Kerfourn (1974); 4. Ters (1973); 5. Greensmith and Tucker (1973); 6. Jelgersma (1961). The limits of the transgression sequence identified by Ters, Jelgersma and Devoy are shown at the base. Curve 3 (Morzadec-Kerfourn) represents data from estuaries of northwest France.

C14 dates of 8640 ± 145 to 8460 ± 145 years bp for levels between −12.7 m to −13.7 m O.D. (−41.7 to −45 ft.) respectively. Biogenic sediment at −3.1 m to −4.6 m O.D. (−10.2 to −15 ft.) gave a date of 3460 ± 100 years bp. These heights and dates agree generally with those from Tilbury and from the Thames estuary model.

Analysis of the biostratigraphy and relative sea-level movements established for the Thames, allows a series of conclusions to be drawn for the form and pattern of environmental change. In the outer estuary the influence of the early Flandrian relative sea-level rise is recorded in the deposition of calcareous clays, silts and sands at depths between −30 m to −34 m O.D. (−98.4–111.5 ft.), prior to 8900 to 8600 years bp. Dominance by the clay-sized fraction at the base coarsening upwards to sand, particularly in the Essex area, may be attributed to the early rapid rate of sea-level rise and the subsequent energy changes taking place in a shallowing, exposed coastal area (Greensmith and Tucker 1971b). The close of biogenic accumulation at ~8200 years bp (Isle of Grain) probably saw the coastal position for M.H.W.S.T. delimited by approximately the −25.5 m (−83.7 ft.) contour. Borehole and geophysical data from the outer estuary shows the rivers Medway and Thames flowing in their present north-east and easterly directions. The rivers follow a series of separate channels joined by a number of short tributary streams. The channel courses remain closely aligned with those of the underlying buried channel phases, although isolated, abandoned sections of the latter have been identified from the Grain/Medway area. Evidence from the Isle of Grain for the vegetational composition of this incipient estuary stage shows a dominance of tree species in the pollen record. Arboreal pollen (AP) values reach >70% total pollen (TP) at this time. *Corylus* type, *Quercus, Pinus* and *Ulmus* form the principal tree pollen taxa, with *Betula* and *Salix* pollen recording low frequencies. Large areas of the now submerged estuary were probably vegetated by local oak-hazel woodland. Pine, elm and oak may have formed more important regional elements in the Thames environment, perhaps growing in the main upon the drier valley sides and London Clay upland surfaces. The presence of high values >20% total land pollen (TLP) of Gramineae and Cyperaceae, with persistent frequencies of *Typha angustifolia/Sparganium* type, *T. latifolia, Myriophyllum* spp., *Lemna*, Compositae Liguliflorae, *Filipendula, Artemisia*, Rosaceae and Umbelliferae pollen, shows the presence of eutrophic fen and aquatic conditions. The pollen spectra suggest a wet or even waterlogged environment with areas of reedswamp and sedge fen close to the river channels. Occasional finds of *Alnus* pollen allows identification of the empirical limit ot *Alnus* at ~8400 years bp.

The waterlogging, related to the rising sea-level, had an important consequence for the local vegetation composition. At Tilbury alder was found growing *in situ* upon the basal Devensian sand surface. Radiocarbon dating of the junction with the overlying wood rich gyttja places the onset of this growth at 8170 ± 110 years bp. Despite the late Boreal VIc pollen assemblage for the deposit, *Alnus* had become a significant contributor to the local pollen rain, values often exceeding 20% TLP (fig.60). This date for the expansion of *Alnus* appears to be the earliest established for its rational limit in south-east, central southern England and East Anglia (Smith and Pilcher 1973). As such it reflects the influence of these local physiographic and edaphic factors, with wet valley conditions favouring the early post-glacial growth of this taxon.

The established pattern of a rapid early relative sea-level movement shows M.H.W.S.T. levels rising at an initial rate of ~1.3 cm/yr (0.51 in.) in the Thames estuary. Inundation had overtaken most areas below the −13/14 m O.D. (−42.6/46 ft.) contour by ~7800 years bp. The nature of the marine transgression (Th I) and the resultant form of coastal change is evidenced by the diatom assemblage at Tilbury (fig. 61). Dominance of mainly brackish water benthonic taxa occurs at the junction between Th I and Tilbury I. These reach a short-lived maximum of >60% total diatoms (TD), with *Cyclotella striata, Nitzschia navicularis, Nitz. punctata, Campylodiscus echeneis, Synedra tabulata* var. *fasciculata* and *Navicula rostellata* forming the principal species. The assemblage suggests a short period of possibly intertidal conditions, represented in the form of shifting mudflat areas. Such an interpretation is supported by the large number of broken and degraded valves, reaching >55% TD. The stratigraphic and pollen evidence shows a similar intertidal phase, indicating the growth of local saltmarsh plant communities. The initial brackish water influence is rapidly replaced after −12.7 m O.D. (−41.7 ft.) by an expansion of marine taxa, which subsequently dominate the diatom flora. These Polyhalobion species reach a maximum of 66% TD and are characterised by *Nitzschia granulata, Melosira sulcata, Cymatosira belgica, Rhaphoneis minor, Rhaph. surirella, Rhaph. amphiceros, Cocconeis scutellum* and *Coscinodiscus excentricus*. These analyses suggest that the tidal head of the estuary had probably reached Tilbury by ~7700 years bp, if not further upstream. Little direct biostratigraphic data exist for a further regression during this time. However, despite the lack of evidence for marine sedimentation beneath the basal gytta at Tilbury (T I), the growth of this deposit may mark a short sea-level regression. Data from Ridge Marsh and Foulness Island in the outer estuary may be used to support this argument. Here a 'peat' rich clay/silt overlies a strongly overconsolidated horizon in brackish/marine sediments at −18.3 m O.D. (−60 ft.), indicating a marine regression (Greensmith and Tucker 1973). A C14 date of 7516 ± 250 years bp upon the level gives a maximal time for this phase, allowing correlation with the biogenic deposit of similar age at Tilbury, although direct lithostratigraphic and chronological comparison is difficult. Following Th I there appears to have been a removal of the marine influence from the inner estuary for ~300 to 400 years, between 7000 to 6600 years bp, to levels below the −9 m O.D. (−29.6 ft.) contour. This left recently submerged areas available once again for plant colonization. Indication locally for a fall of sea-level during this time is given by the *in situ* growth of freshwater oak-alder wood peat. A return to brackish water conditions above Tilbury II, prior to a further expansion of marine diatom taxa (fig. 61), supports this arguments. Pollen and macro-fossil analyses of the peat show a domination of the vegetation initially by alder carr and later by alder-oak fen woodland (fig. 60). The vegetation composition was strongly influenced locally by the growth of *Phragmites* reedswamp and saltmarsh communities, particularly before the eventual extinction of the fen woodland by Th II. Inwashing of *Pinus* pollen, brackish water foraminifera, pre-Quaternary spores and hystrichospheres, together with a rise in the silt sized fraction at the top of the biogenic levels, suggest frequent flooding before final marine inundation.

Thames II continued the submergence of the coastal zone, although the rate of sea-level rise had fallen to ~0.5 cm/yr. (0.197 in.). M.H.W.S.T. levels had probably reached the −5.0 m O.D. (−16.4 ft.) contour in the river area by ~5500 years bp before a further regression occurred (T III). The growth of a thick biogenic deposit throughout the inner estuary and the formation of overconsolidated horizons below Dengie Flats and Maplin Sands evidence a removal of the marine influence, lasting until ~4000 years bp. Changes from east to west in the composition of the biogenic deposits show the establishment of differing environmental conditions within the estuary during this time. In the Tilbury area, *Phragmites* reedswamp and saltmarsh peats dominate the deposits, characterized by high frequencies of Gramineae, Cyperaceae, *Typha angustifolia/Sparganium* type, Chenopodiaceae and Compositae *Bellis* type pollen (fig. 60). Brackish water conditions probably continued here, as supported by the presence of brackish water foraminifera, pre-Quaternary spores and hystrichospheres throughout. It is likely that the exposed coast of the outer estuary remained in close association with the full marine influence to prevent the preservation of any organic sequences formed, sea-level possibly falling to only the −7.5 m O.D. (−24.6 ft.) contour.

Upstream, at sites west of Broadness Marsh, an initial phase of *Phragmites* and saltmarsh peat accumulation occurs. This is overtaken by the development of an alder-oak fen wood peat, before final return to *Phragmites* reedswamp and saltmarsh deposits. Pollen, macro-fossil and stratigraphic analyses show a series of seral vegetational changes taking place (Devoy 1979) (fig. 62). Sedge fen, reedswamp and saltmarsh communities are replaced upwards by the growth of freshwater alder carr. Later expansion of *Corylus* type and Quercus as the dominant tree pollen taxa indicates development of drier oak-hazel fen woodland. *Filicales* and *Thelypteris palustris*, favouring the shady and damp woodland conditions, show high spore frequencies expanding to >50% TLP + Pteridophytes and support this interpretation. The corroded and degraded state of some of the pollen at this stage indicates probable periodic oxidation of the peat under the influence of a fluctuating water table and the continuance of a low sea-level. These conditions would favour the growth of oak-hazel fen wood. The prominence of Gramineae, Cyperaceae and non-arboreal pollen (NAP) shows that this environment probably remained seasonally wet, allowing the continued growth of light-demanding, herb-dominated plant communities (fig. 62).

The impact of man upon the estuary environment, frequently affected by changes in the position of the coastal zone and related hydrology, is difficult to gauge. From the first removal of the marine influence of Th II, man was apparently present in the area. Evidence comes from pottery finds of probably early Neolithic age, lying *in situ* upon the surface of estuarine sand/silt at Ebbsfleet (Burchell and Piggott 1939). A C14 assay taken at the base of the overlying peat T III, dates the onset of biogenic accumulation here to 4660 ± 110 years bp (Barker and Mackay 1963). The continued presence of early Neolithic and later cultures in the emerging reedswamp and fen woodland zones, perhaps representing hunter/collector activities, is demonstrated by the numerous artifacts found in association with T III. In Plumstead, Erith, Beckton and Dagenham Marshes flint microliths, arrowheads, handaxes and temporary hearths, assigned broadly to the Neolithic and early Bronze Age periods, have been recorded from many different levels in these peats (Spurrell 1889; Gilbert 1930). Evidence of earlier habitation of the coastal zone itself comes from sites in the outer estuary. At Hullbridge on the river Crouch and at Lower Halstow, finds from similar late Mesolithic cultures were found in contact with the surface of estuarine sandy clays (Burchell 1925–27, 1957; Akeroyd 1972). Both sites directly underlay biogenic deposits dated by pollen analysis to pollen assemblage zone (P.Z.) VIIa. At Lower Halstow, high Chenopodiaceae pollen frequencies with other saltmarsh taxa from the clays and basal peat layers, support the evidence of occupation in an intertidal or high tidal environment. The close stratigraphic association of later Neolithic artifacts and temporary hearths with both these sites indicates a long history of human activity in these estuary areas. However, in the light of the present Thames sea-level data the significance of these sites for coastal change is open to re-interpretation, due to possible inaccuracies in their dating and relationship to O.D.

Man's influence upon the estuary's environment and vegetation at this time may be seen as purely local, with perhaps occasional clearance and probably isolated burning of the fen woods. There appears to be little direct indication of any prolonged habitation of these wet lowland areas. Progressively as pastoral and arable economies became more important, the drier chalk uplands would have become centres for occupation, the estuary serving as a seasonal and supplementary source of fresh food. Major changes in the regional vegetation composition, reflected in the pollen and macro-fossil analyses of the T III sediments, support this interpretation although the evidence is somewhat equivocal. The strong local pollen influences of the Thames reedswamp and fen woodland communities suppress the more regional pollen elements.

A decline in *Ulmus* pollen values forms a distinctive feature in most pollen diagrams for north-west Europe covering P.Z. VIIa/b (Godwin 1975). This may arguably be attributed to Neolithic agricultural practices or to a combination of climatic, edaphic and anthropogenic factors (Ten Hove 1968; Sims 1973). It appears consistently in the Thames pollen spectra, occurring approximately at the onset of T III biogenic accumulation. A C14 assay at Stone Marsh gives a date of 4930 ± 110 years bp for the elm decline. Radiocarbon dates from other sites, taken at the base of T III but close to the recorded feature, are conformable with this date showing the decline to have occurred at ~5000 years bp. Its appearance in the stratigraphy with the first archaeological indicators of man's presence is significant, possibly showing the initial stages in man-induced vegetational change. Identification of the feature confirms the maximal age for the artifacts and occupation levels found in the inner estuary. Recognition of the elm decline from the Thames sites is based upon the fall of *Ulmus* pollen frequencies from 5/4.5% to <1% TLP (figs. 60 and 61). This differential appears low as a basis for de-limiting the decline, but is consistent with similar *Ulmus* pollen values for other parts of eastern England (Devoy 1979). The influence of *Quercus, Alnus* and *Corylus* type pollen, representing important *in situ* constituents of the Thames vegetation, may locally depress the *Ulmus* frequencies at the sites studied. *Plantago lanceolata* pollen makes its first significant appearance at this time, in conjunction with a rise in ruderal pollen taxa. A low but persistent increase in *Fraxinus* pollen frequencies and a sharp rise in *Pteridium aquilinum* values accompany these changes. However, as evidenced at Broadness Marsh (fig. 61) the behaviour of these supporting taxa is not uniform for each of the Thames sites. Variations in the implied vegetation trends may in part be dependent upon local hydrological changes, resulting from the contemporary sea-level movements. The growth of sedge fen and reedswamp vegetation at the beginning of T III, prior to the recorded elm decline, may alternatively account for the expansion of these light-demanding pollen & spove taxa. This may be particularly true for sites toward the seaward end of the estuary. Closer association here with continued estuarine/marine conditions would have favoured the longevity of open plant communities.

Changes in *Tilia* values and associated pollen taxa in the Thames estuary, may evidence both a continuance and expansion of the anthropogenic influence upon vegetational composition. *Tilia* pollen frequencies in southern and eastern England often attain values >20% total tree pollen during P.Z. VIIa and it may have formed an important constituent of the vegetation at this time (Godwin 1975; Birks, Deacon and Peglar 1975). At the Thames sites the behaviour of *Tilia* pollen supports this regional picture. Values remain high though sporadic in form during P.Z. VIIb, generally falling between 6 to 11% TLP but occasionally rising to

>15% TLP. Such high frequencies and the erratic form of the curves suggest that the taxon may have grown locally in the estuary's fen woods and upon the nearby Chalk slopes (Devoy 1979). The decline to low and sporadic values <1% TLP later in P.Z. VIIb coincides with the major phase of recorded woodland recession (figs. 60 and 62). This is coupled with a large expansion of NAP to values >45% TP. Cereal type pollen makes its first appearance at this time, together with an increase in *Pteridium aquilinum* spore values and an influx of pollen from calcicole taxa, particularly *Poterium* and Rubiaceae pollen. *Fagus* and *Acer* pollen, probably *Acer campestre*, also occur in significant frequencies. These changes suggest that the accompanying apparent *Tilia* decline may be associated with a major expansion of man's farming activities (Turner 1962, 1965, 1970; Mittre 1971), centred upon clearances of the coastal Chalk upland. However, the decline in *Tilia* pollen occurs with the onset of the marine transgression Th III in the area at ~4000 years bp, although the decline at Broadness and Stone Marshes occurs before the final inundation. The resultant increase in wetness and changed edaphic conditions would not have favoured the growth of *Tilia* locally in the estuary, and may equally well account for the observed fall of values. The major expansion in NAP frequencies appears in great part due to the generation of sedge fen, reedswamp and saltmarsh communities in the increasingly water-logged environment. The frequent finds of seeds from *Aster tripolium, Juncus gerardii* and *Juncus maritimus* in these transitional biogenic to inorganic levels indicates growth close to M.H.W.S.T. The high Gramineae pollen values are probably due to the wider growth of *Phragmites australis*. Macro-fossil remains of this taxon are abundant at these levels in the form of the characteristic broad yellow monocot stems and leaves.

The return of brackish/marine water conditions shortly after ~4000 years bp resulted in an initial phase of erosion and re-working of the T III surface. Broken and degraded valves of mainly brackish water-dominated diatoms indicate the development of broad estuarine mudflat zones, with the entrenchment of intertidal creeks through the underlying peats. Subsequent changes in coastal positions are evidenced in at least two further regression phases in the inner estuary, T IV and V. These occur in the form of thin, non-persistent, silty monocot peats intercalated with brackish/marine clays and silts. The timing and height of the events varies, particularly T V, and may result from different rates of sedimentation and relative subsidence trends (Devoy 1977). Pollen and macro-fossil analyses show the growth of saltmarsh- and reedswamp-dominated plant communities at the seaward end of the estuary (fig. 60). Continued high water levels are indicated by the large silt-sized fraction and the presence of brackish water foraminifera, pre-Quaternary spores and hystrichospheres. Upstream this vegetation gives way to freshwater reedswamp, with shortlived phases of oak, hazel and willow fen wood growth. However, the woodland does not recover its former regional or local dominance, remaining at levels <23% TP. This feature, in conjunction with identification of cereal type pollen to levels of 1% TLP and the appearance of associated ruderal taxa, may show the permanence of man's impact upon the vegetation by this time.

The continued direct use by man of these coastal environments can be judged from artifact finds associated with the surface of T III and the deposits of T IV. The habitation phases of the Lyonesse surface in the outer estuary (Warren 1932;

Akeroyd 1966) shows persistent occupation from ~4700 years bp through to inundation of the area by ~3800 years bp, evidencing the attraction of the coastal zone. The surface may tentatively be correlated with T III on stratigraphic, archaeological and C14 data. In the inner estuary temporary occupation sites from T III and T IV have been dated archaeologically to the Bronze Age (Spurrell 1889; Gilbert 1930). This dating is supported by the pollen and C14 evidence. Dug-out canoes found at the base of clay/silt filled channels cut through T III have also been attributed to this period (Whitaker 1889). However, the anomalous stratigraphic position of the canoes and the lack of suitable pollen material, makes independent dating of these difficult in the absence of radiometric data. Similarly the finding of the Tilbury human skeleton (BM No. 1913) upon a sand ridge of late Devensian age at −9.44 m O.D. (−31 ft.) and overlain by estuarine clay/silt has caused much controversy. Pollen analysis by Churchill (1963) of scrapings from the medulla of the right femur indicates that this was buried during P.Z. VIIb, contradicting its stratigraphic position.

This evidence suggests that the estuary zones remained for Bronze Age man areas of hunting, fishing and gathering. In such a context the coastal zone may be seen to function ecologically as an ecotone. The different marine, freshwater and terrestrial habitats combine to support varied and inter-connected plant and animal communities, providing a rich and highly productive food source. The successive changes in coastal position would have caused fluctuations in the extent and diversity of these communities both spatially and through time. For man, the richest and most productive environments would have been presented during times of sea-level regression and expansion of the coastal zone.

In the biogenic deposits of each of the Thames regressions, macro-fossil remains of *Corylus* support the pollen evidence for the importance of hazel in the local vegetation composition. The use of hazel by man for both food and materials is well documented (Clark 1954; Godwin 1975) and it is probable that hazel, together with other edible fruits from the fen woods, was used. Interestingly, from the T III peats at Littlebrook, cut and shaped branches of *Alnus, Corylus* and *Quercus* were recorded and it is in this layer that the elm decline occurs. However, an animal or natural cause cannot be excluded. The coastal vegetation supported a rich animal population. Records of their remains (Spurrell 1889) show that *Castor fiber, Sus scrofa, Bos* sp., *Cervus elaphus, Canis lupus, Arvicola* spp. and *Sciurus vulgaris* were common in the T II and T III levels. In the reedswamp and sedge fen areas *Phragmites australis*, with other grasses and sedges, may have been gathered for domestic purposes, whilst providing a habitat for wild fowl. Abundant shell material indicates the existance of quiet open water pools in the Plumstead, Erith and Dartford Marsh areas, close to the Chalk upland (Shillitoe 1958). In these freshwater zones of the fen wood environment the edible freshwater mussels *Anodonta* and *Unio* were common. During marine transgressions, particularly, the intertidal area presented a productive salt-marsh and mudflat ecosystem. Chenopodiaceae, *Salicornia* spp. and other edible plants occurred here, in an environment naturally suited to livestock grazing as well as to providing a source of crustaceans and fish. Throughout the Flandrian intertidal sequences, shell remains evidence the abundance of *Cardium edule, Mytilus edulis, Ostrea edulis, Littorina littorea* and *Buccinum* sp. (Greensmith and Tucker 1973). Conclusive proof for the use of each of these plant communities by man in the Thames is lacking, although the vegetational and archaeological data would support this idea.

Assuming the operation of the present tidal regime (Admiralty Tide Tables 1977), the effective limit of M.H.W.S.T. between the outer and inner estuary by ~1750 years bp, probably lay −2 to −3 m (−6.56 to −9.84 ft.) respectively below today's levels. M.H.W.S.T. at Tilbury occurred at +0.4 m O.D. (1.3 ft.), fixed by the height of the transgressive contact for T V. Equated with the Roman period, this time evidences a major phase of building and habitation in the coastal zone. Archaeological material (Evans 1953; Kent 1960; Merrifield 1965; Akeroyd 1966) documents a relatively dense level of occupation along the Thames shores and even upon the exposed peat surface during the regression T V. The subsequent large rise of M.H.W.S.T. levels may in part have been due to the increased definition of estuary shape, following embanking and removal of the natural flood depressions. The Romans probably began this process, although little evidence of their embankments now exists outside central London (Spurrell 1885; Evans 1953). It is more likely that the need for protection of the occupation areas from flooding in the third century A.D. (Akeroyd 1966) resulted not only from the effects of such man-made changes but also from the onset of a further marine transgression Th V. Indication that relative sea-level had continued to rise, following a regression of ~150 years during T V, comes from analysis of the diatom assemblages (fig. 61). A sharp decline in marine taxa to <10% TD is recorded in Th IV, as brackish water species become dominant. The rise in the numbers of broken valves indicates re-working of the sediments in an emerging intertidal environment. At the upper contact a return of marine taxa to >25% TD and a decline in the frequency of broken valves shows a resumption of the marine tidal influence. The Polyhalobion species are dominated by *Cymatosira belgica, Melosira sulcata, Rhaphoneis minor* and *Rhaph. surirella*.

A feature of these last transgressions, however, is the progressive decline in the marine diatom influence. Oligohalobion and indifferent species reach significant values in the assemblages for the first time. The species *Fragilaria construens* var. *venter, Fragilaria brevistriata, Fragilaria pinnata, Navicula cryptocephalla* and *Navicula cincta* are of importance particularly in the beginning of Th V. Mesohalobion taxa, represented by *Diploneis didyma, Diploneis suborbicularis, Navicula peregrina, Nitzschia punctata* and *Scolopleura brunksii* dominate the assemblages overall. This pattern shows a markedly reduced marine tidal influence toward the present day and the increasing importance of freshwater to salt-tolerant benthonic and epiphytic taxa. The diatoms indicate the formation of shallow-water middle estuarine conditions and the existence of saltmarsh in the inner estuary (Aleem 1950a, b; Round 1960; Hendey 1964; Hodson and West 1972). Such apparent freshening of the estuarine environment may evidence the decline in the rate of relative sea-level rise, coupled with the increasing sediment heights and rates of accretion. These factors would effectively reduce the marine tidal influence upon the area. Alternatively, climatic data may explain this feature. The onset of the freshening from the beginning of Th IV occurs at ~2900 years bp. It coincides with a time of increased storminess and a drop in average temperatures by 2°C over Europe from ~2950 years bp (Lamb 1977). The probable resultant increased rainfall and flow of freshwater into the estuary may be shown by these changing diatom assemblages, although a local ecological explanation cannot be excluded. Similarly, the abandonment of mid third-century A.D. occupation levels in the southern Fenlands (Churchill 1970) and the Thames due to flooding may be more the result of these climatic changes than a consequence of a rising sea-level.

The pattern of coastal and environmental evolution in the estuary in post-glacial time can be seen as a function of long term relative sea-level movements. The trend since medieval times, and more importantly in recent decades, of continuing inundation is most easily correlated with the process of embanking and channel alteration. The natural parameters of a rising relative sea-level and changes in estuary salinity are probably not of great importance for the immediate future. The influence of changes in wind and air pressure, as evidenced in storm surges, do have an influence on the coastal zone (Lamb 1969; Prandle 1975). However, although underlining the danger to man of changes in coastal position, such phenomena are not causative factors. In perspective, observed increases in tidal amplitude owe more to man than to natural causes and must be regarded as an important additional factor to the long term trends of sea-level movement. Historically, coastal alterations have been beneficial to man, offering a rich natural zone for exploitation. Ironically, today the location of urban structures in the estuary area, with all its economic advantages, is threatened by this very process of coastal change.

## BIBLIOGRAPHY

Admiralty, 1977. *Admiralty Tide Tables; European Waters including the Mediterranean Sea* (vol. 1). London: Hydrographer of the Navy.

Akeroyd, A. V. 1966. Changes in relative land- and sea-level during the post-glacial in southern Britain with particular reference to the post-Mesolithic period. Unpublished M. A. thesis, University of London.

—— 1972. Archaeological and historical evidence for subsidence in southern Britain. *Phil. Trans. R. Soc. Lond.* A. **272**, 151–69.

Aleem. A. A. 1950a. Distribution and ecology of British marine littoral diatoms. *J. Ecol.*, **38**, 75.

—— 1950b. The diatom community inhabiting the mud flats at Whitstable. *New Phytol.*, **49**, 174–88.

Baker, H. and Mackay, J. 1963. British Museum natural radiocarbon measurements IV. *Radiocarbon,* **5**, 104–8.

Birks, H. J. B., Deacon, J. and Peglar, S. 1975. Pollen maps for the British Isles 5000 years ago. *Proc. R. Soc. Lond.* B. **189**, 87–105.

Burchell, J. P. T. 1925–27. The shell mound industry of Denmark as represented at Lower Halstow. *Proc. Prehist. Soc. E. Anglia,* **5**, 73–8, 217–221, 289–96.

—— 1957. The Upchurch Marshes. *Kent Arch. Newsletter,* **6** (4) and 89–91.

—— and Piggott, S. 1939. Decorated prehistoric pottery from the bed of the Ebbsfleet, Northfleet, Kent. *Antiq. Journ.*, **14** (4), 405–20.

Carr, A. P. and Baker, R. E. 1968. Ordord, Suffolk: evidence for the evolution of the area during the Quaternary. *Trans. Inst. Brit. Geogr.*, **45**, 107–23.

Churchill. D. M. 1963. A preliminary report upon the Tilbury human skelton, B.M. No. 1913. Unpublished paper. Godwin Collection. Botany School, Cambridge.

—— 1970. The Neolithic to Roman stratigraphy of some sites in the south Fens in relation to land and sea-level. Unpublished paper, Godwin Collection, Botany School, Cambridge.

Clark, J. G. D. 1954. *Excavations at Star Carr.* Cambridge: University Press.

Devoy, R. J. H. 1977. Flandrian sea-level changes in the Thames estuary and the implications for land subsidence in England and Wales. *Nature* **270** (5639), 712–15.

—— 1979. Flandrian sea-level changes and vegetational history of the Lower Thames estuary. *Phil. Trans. R. Soc. Lond.* B. **285**, 355–407.

Evans, J. H. 1953. Archaeological horizons in the north Kent marshes. *Arch. Cant.* **64,** 103–46.

Gilbert, C. J. 1930. Land oscillations during the closing stages of the Neolithic depression. In K. S. Sandford (ed.), *Paper 2, 2nd Rep. Comm. Pliocene and Pleistocene terraces. Int. Geogr. Union,* 93–101.

Godwin, H. 1975. *The History of the British Flora.* 2nd edn. Cambridge: University Press

Greensmith, J. T. and Tucker, E. V. 1971a. Overconsolidation in some fine grained sediments; its nature, genesis and value in interpreting the history of certain English Quaternary deposits. *Geol. en Mijnb.* **50** (6), 743–8.

—— 1971b. The effects of late Pleistocene and Holocene sea-level changes in the vicinity of the river Crouch, East Essex. *Proc. Geol. Assoc.,* **82** (3), 301–22.

—— 1973. Holocene transgressions and regressions on the Essex coast outer Thames estuary. *Geol. en Mijnb.* **52** (4), 193–202.

Grieve, H. 1959. *The Great Tide: the Story of the 1953 Flood Disaster in Essex.* London: Essex County Council.

Hendey, N. I. 1964. An introductory account of the smaller algae of British coastal waters. Part V. Bacillariophyceae (Diatoms). *Fishery Investigations* Ser. IV. London: H.M.S.O.

Hodson, F. and West, I. M. 1972. Holocene deposits of Fawley, Hampshire, and the development of Southampton Water. *Proc. Geol. Assoc.,* **83,** 421–41.

Horner, R. W. 1972. Current proposals for the Thames barrier and the organisation of the investigations. In K. C. Duntam and D. A. Gray (eds.). A discussion on problems associated with the subsidence of S. E. England. *Phil. Trans. R. Soc. Lond.* A. **272,** 179–86.

Jelgersma, S. 1961. Holocene sea-level changes in the Netherlands. *Meded. Geol. Sticht.* C VI, **7,** 1–100.

Kent, J. P. C. 1960. *The Archaeologist in Essex, Hertfordshire, London and Middlesex in 1959.* C.B.A. Group 10, 1–58.

Lake, R. D., Ellison, R. A., Henson, M. R. and Conway, B. W. 1975. *South Essex Geological and Geotechnical Survey*. Pt. 2. *Geology*. London: National Environment Research Council, Institute of Geological Sciences.

Lamb, H. H. 1969. On the problem of high waves in the North Sea and neighbouring waters and the possible future trend of the atmospheric circulation. Personal communications to H. Godwin.

—— 1977. The late Quaternary history of the climate of the British Isles. In F. W. Shotton (ed.), *British Quaternary Studies: Recent Advances.* Oxford: Clarendon Press.

Merrifield, R. 1965. *The Roman City of London.* London: Benn.

Mittre, V. 1971. Fossil pollen of *Tilia* from the East Anglian Fenland. *New Phytol.,* **70,** 693.

Morzadec-Kerfourn, M. T. 1974. Variations de la ligne de rivage Armoricaine au Quaternaire. *Mém. Soc. géol. mineral. Bretagne.,* **17,** 1–208.

Prandle, D. 1975. Storm surges in the southern North Sea and river Thames. *Proc. R. Soc. Lond.* A. **344,** 509–39.

Rossiter, J. R. 1972. Sea-level observations and their secular variation. *Phil. Trans. R. Soc. Lond.* A. **272,** 131–9.

Round, F. E. 1960. The diatom flora of a saltmarsh on the river Dee. *New Phytol.,* **59.** 332–48.

Shillitoe, J. S. 1958. Holocene mollusca from Plumstead Marshes. *Lond. Nat.,* **38,** 23–4.

Sims, R. E. 1973. The anthropogenic factor in East Anglian vegetational history: an approach using A.P.F. techniques. In H. J. B. Birks and R. G. West (eds.), *Quaternary Plant Ecology.* Oxford: Blackwell.

Smith, A. G. and Pilcher, J. R. 1973. Radiocarbon dates and vegetational history of the British Isles. *New Phytol.,* **72,** 903–14.

Spurrell, F. C. J. 1885. Early embankments on the margins of the Thames estuary. *Arch. Journ.* **42,** 269–302.

—— 1889. On the estuary of the Thames and its alluvium. *Proc. Geol. Soc.,* **11,** 210–30.

Ten Hove, H. A. 1968. The *Ulmus* fall at the transition Atlanticum – sub-Boreal in pollen diagrams. *Palaeogeog. Palaeoclimatol. Palaeoecol.,* **5,** 359–69.

Ters, M. 1973. Les variations du niveau marin depuis 10,000 ans le long du littoral atlantique francais. In *La Quaternaire: géodynamique, stratigraphique et environnement.* Christchurch, N.Z.: Congress International de l'Inqua, 114–35.

Turner, J. 1962. The *Tilia* decline: an anthropogenic interpretation. *New Phytol.*, **61,** 328–41.
—— 1965. A contribution to the history of forest clearance. *Proc. R. Soc. Lond.* B. **161,** 343–54.
—— 1970. Post Neolithic disturbance of British vegetation. In Walker, D. and West, R. G. (eds.), *Studies in the Vegetational History of the British Isles.* Cambridge; University Press.
Warren, S. H. 1932. The Palaeolithic industries of the Clactonian and Dovercourt districts. *Essex Naturalist,* **24,** 1–29.
Whitaker, W. 1889. *Geology of London.* Mem. Geol. Surv. **1,** 454–77.

# *Index*

Compiled by PHOEBE M. PROCTER


Aartswoud (Netherlands), beaker settlement at, 126
Alblasserwaard (Netherlands), Bell Beaker occupation in, 112
Aldeburgh (Suffolk), 137–8
Akeroyd, A. V., cited, 1, 21, 40
Aberdeen (Grampian), 17, 21
Aberweythel (Gwent), medieval mill at, 34
Algarkirk (Lincs.), 61
Alt estuary (Lancs.), 74, 79, 84, 91
Amsterdam (Netherlands), tide gauge at, 12
Ancaster (Lincs.), 69, 71
Anglian Water Authority, 58, 65
Anglo-Saxon Chronicle, cited, 48
Antarctic ice sheets, 7, 8
Appledore (Kent), 49, 52
'Archaeology and Coastal Change in the Netherlands', by Dr. L. P. Louwe Kooijmans, 106–33; bibliography 131–3; 'Archaeology and Coastal Change in the North-West'. by Prof. G. D. B. Jones, 87–102; bibliography, 102
Ashmead, P., cited, 83
Askew, G. P., *see* Green, R. D.
Aslackby Fen (Lincs.), 64
Atlantic Ocean, 6, 8, 10
Atmospheric pressure, 10, 15

Bain, river, 71
Balchin, W. G. V., *see* Lewis, W. V.
Barrows, Bronze Age, 69
Beckett, Dr. S. C., cited, 105
Beckfort (Cumbria), 91
Bergschenhoek (Netherlands), Neolithic settlement at, 112, 118
Betuwe (Netherlands), Roman period occupation at, 112
Binney, E. W., and Talbot, J. H., cited, 75, 85
Birkdale Hills (Lancs.), 74
Blackpool (Lancs.), 83
Black Sluice Internal Drainage Board, 58, 64
Boon, G. C., 'Caerleon and the Gwent Levels in Early Historic Times', by, 24–36
Bourne (Lincs.), 56, 69
Bourne-Morton Canal (Lincs.), 59, 64
Bovenkarspel (Netherlands), Bronze Age site, 124–5
Branigan, K., cited, 27
Brede, river, 39, 43, 45, 48
Briquetage, 44, 61, 64
Broadness Marsh (Kent), 81, 143
Brookland (Kent), 47
Burgh-le-Marsh (Lincs.), 61, 69
Burials: Neolithic, 116, 121; Roman (cremation), 44; *see also* Barrows

Caerleon (Gwent) and Gwent Levels in early historic times, 24–36; departure of 2nd Augustan Legion from, 28, 30; Roman quay, 24–5, 29–30; *see also* Gwent Levels
Caerwent (Gwent), Silurian capital, 28

Caldicot and Wentlooge Levels Drainage Board, 19th-century surveys of Gwent Levels in offices of, 32
Camden, William, cited, 32, 51
Canvey Island (Essex), 136–7
Cardiff (S. Glamorgan), Taff-Ely moor at, 33
Car Dyke (Lincs.), 61, 64, 67, 71
Cartmel (Lancs.), Kirkhead Cavern in, 83
Catuvellauni, 69
Celsius, cited, 12
Channel Islands, 16
Chapel Hill (Lincs,), 67, 71
Churchill, D. M., cited, 144
Clark, J. A., *see* Farrell, W. E.
Clason, Dr. A. T., cited, 119
Clevedon (Avon), Iron Age and Roman finds at, 25, 26, 27
Cleveleys (Lancs.), 78
Cliffe (Kent), 136–7
Climatic change, 10, 124, 142, 145, 146
Commission for Recent Crustal Movements, 17
Committee on Tidal Gauges, 17
Compaction, 2, 15, 45, 67, 126
Cooling (Kent), 136–7
Coriolis effect, 8, 10, 15
Coritani, 69
Cornwall, 12; *see also* Newlyn
Crustal movements, 3, 8, 10, 12, 16, 21
Cumbria; coastal change in, 74; ancient settlement in, 89; milefortlets in, 91
Cunliffe, Prof. B. W., 'The Evolution of Romney Marsh: a Preliminary Statement', by, 37–55
Currie, R. G., cited, 7

Darwin Rise, 8
Dengie Flats (Essex), 137
Denton, G. H., and Karlen, W., cited, 7
de Rance, C. E., cited, 85
Devon. 12
Devoy, Dr. R., 'Post-glacial Environmental Change and Man in the Thames Estuary: a Synopsis', by, 134–48
d'Hieres, G. C., and Le Provost, C., cited, 10
Diatom assemblages, 145
Dowker, G., cited, 42
Downholland Cross (Lancs.), ancient settlement at, 91
Downholland Moss (Lancs.), 75, 76, 78, 79, 80, 84, 91
Drainage engineering: Roman, 28, 64, 71; Medieval, 32–4, 37, 39, 40, 51–2, 54, 64, 113, 131; Post-medieval, 40, 43, 58, 64, 67
Drigg (Cumbria), 74
Dugout canoes, 89, 144
Dunbar (Lothian), 21
Dune formation, 74, 76, 79, 107, 112, 122, 123–4, 131
Dungeness (Kent), 37, 38, 39, 42, 50, 52
Dymchurch Wall (Kent), 40, 43, 44, 52

East Anglia: 10, 81, 134; *see also* Felixstowe; Sudbury; Thames estuary
Eastbridge (Kent), 47
Ebbsfleet (Kent), Neolithic occupation at, 141
Ehenside Tarn (Cumbria), 89
Elliott, James, cited, 37, 43
English Channel, 10, 12, 15
Ente, P. J., cited, 124
Eskmeals (Cumbria), 74
Eustatic change, 2, 3, 8, 12, 16, 21, 76, 87, 136
Everard, C. E., 'On Sea-Level Changes', by, 1–23

Fairbridge, Prof. R. W., cited, 80
Fairfield (Kent), 47
Farming, 26, 28–9, 30, 32, 33, 34, 116, 118, 119, 121, 123, 126, 127, 142, 143
Farrell, W. E., and Clark, J. A., cited, 8
Felixstowe (Suffolk), 21
Fennoscandian sheet, 8, 12
Fishing, 116, 118, 119
Fishtoft (Lincs.). 61
Flandrian transgression, 7, 78, 81, 83, 84, 85, 136, 137, 138, 144
Fleet (Lincs.), 61
Forests, 42–3, 54, 78, 84, 87, 89, 116, 123
Formby (Lancs.), 79, 84
Foulness (Essex), 137, 140
Friesland (Netherlands), Iron Age settlements in. 110, 112, 127–9
Fylde (Lancs.): 75, 78, 79, 81, 83, 100, 102; ancient settlement in, 91

Galloway, mesolithic settlements in, 89
Gedney (Lincs.), 61
Geoid, *see under* Sea level changes
George, K. J., and Thomas, D. K., cited, 15
Geyhl, M. A., and Streif, H., cited,
Gilbert, C. J., cited, 38
Gilbert de Clare, cited, 30
Godwin, Sir Harry, cited, 24, 25, 84
Goldcliff (Gwent): Roman inscribed stone from, 27–28; priory, land reclamation and erosion at, 33–4; 1606 flood plaque in church, 30
Gordon, D. L. and Suthons, C. T., cited, 3
Green, R. D., cited, 42, 43, 45, 47, 48, 49, 51
Green, R. D., and Askew, G. P., cited, 47
Greenland ice sheets, 7
Gresswell, R. K., cited, 75, 76, 85
Groningen (Netherlands), Iron Age occupation of, 110, 112, 127–9
Gulf of Bothnia, 12
Gwent Levels: 24–34; drainage of, Roman, 28, medieval, 32–4; finds from, Iron Age and Roman, 26, 29, medieval, 30, 32; marine transgression in, late Roman, 24, 27, medieval, 30, 32, 33, 1606–7 30, 32; pastoral farming in, Roman, 28–9, medieval, 30, 32, 33, 34
Gwynllyw, St.: 31–2; church of (St. Woolo's), 32

Hacconby Fen (Lincs.), Roman saltern site, 58–9, 61, 65
Halkyn (Clwyd), Roman lead mines at, 97, 99
Halstow, Lower (Kent), 134, 141
Hassall, Charles, cited, 28
Hawkins, A., cited, 134
Hawkins, A. B., cited, 25, 26
Haynes, Dr. John, cited, 30
Hazendonk (Netherlands): Mesolithic occupation at, 116; Neolithic settlements at, 112, 118–21
Helpringham Fen (Lincs.), 61, 64, 67
Heyworth, A., *see* Kidson, C.
Hibbert, Dr. F. A., 'Possible Evidence for Sea-Level Change in the Somerset Levels', by, 103–5
Highbridge (Somerset), buried inlet at, 26
'Hillhouse Sea' (Lancs.), 76, 85
Hoinkes, A., cited, 8
Holbeach (Lincs.), 61, 69
Holmes, Rice, cited, 56
Holocene sea-levels, 1, 3, 7, 12, 87, 88, 107, 122
Homan, W. M., cited 38
Horbling Fen (Lincs.), 69
Horncastle (Lincs.), 69, 71
Hullbridge (Essex), Mesolithic occupation at, 141
Hunter–fisher–gatherer communities, 116, 118, 119, 127, 141, 144
Hythe (Kent), 37, 43, 47, 49

Ice fluctuations, 7, 8
Iceni, 69
*Illtyd, St., Life of,* cited, 33
Indian Ocean, 10
International Geological Correlation Programme, 3
'Iron Age and Roman Coasts around the Wash', by Brian Simmons, 56–73; bibliography, 72–3; for details *see* Lincolnshire Fens
Isle of Grain (Kent), 137, 139
Isle of Man, 83
Isle of Wight, 10
Isostatic rebound, 3, 12, 16, 21, 40, 42, 85, 87
Ivychurch (Kent), 47

Jardine, W. G., cited, 3
Jardine, W. G., and Morrison, A., cited, 87, 89
Jelgersma, S., cited, 79, 81
John O'Groats (Highland), 16
Johnson, D., cited, 3
Jones, Prof. G. D. B., 'Archaeology and Coastal Change in the North-West', by, 87–102
Julius Caesar, invasion of Britain by, 38

Karlen, W., *see* Denton, G. H.
Kelsey, J., cited, 17
Kidson, C., and Heyworth, A., cited, 103
King, L. C., cited, 10
Kingsweston (Avon), Roman villa, 26–7, 34
Kyme (Lincs.), 71

Lancashire coast: changes in, 74–85; palaeoenvironments, 76–80, 85;

ancient settlement in, 91, 93; Palaeolithic period, 83; Mesolithic period, 83, 84, 89; Neolithic period, 83–4; Bronze Age period, 84; Iron Age period, 79; Romano-British period, 84; Dark Ages period, 79
Laurentide sheet, 8
Law, C. R., cited, 15
Lead industry, Roman, 95, 97, 98, 99
Le Provost, C., *see* d'Hieres, G. C.
Levelling networks, 16, 20
Lewis, W. V., cited, 38
Lewis, W. V., and Balchin, W. G. V., cited, 38, 47, 50, 52
Limen, river, 47, 48, 49
Lincoln (Lincs.), 69, 71
Lincolnshire Fens, the: geology, 56–7; Bronze Age finds, 69; Iron Age: coastline, 67; salterns, 61, 67; finds, pottery, 61, 65; Roman: arable agriculture, 69; coastline, 65, 67; finds, 58–9, 65, 67; saltern sites, 59, 61, 64, 65, 69; tribal territories, 69; Saxon settlement, 61
Littlebrook (Kent), 144
Lochar Moss (Dumfries and Galloway), 87, 89
Locke, Stephen, cited, 25
Long Sutton (Lincs.), 67
Louwe Kooijmans, Dr. L. P., 'Archaeology and Coastal Change in The Netherlands', by, 106–33
Lundy, 12
Lydd (Kent): 47, 49; Roman occupation evidence near, 45
Lympne (Kent), Roman fort at, 45
Lytham (Lancs.), 79, 81, 84, 85, 100

Magor (Gwent): Roman settlement at, 26; medieval drainage at, 34
Man, impact of on environment, 15, 40, 67, 141–2, 143, 146
Margam abbey (Gwent), land reclamation by, 33
Marine regression, 6, 67, 75, 80, 81, 84, 106, 113, 124, 128, 137, 140, 144
Marine transgression, 6, 24, 27, 30, 32, 33, 40, 42–3, 75, 76, 78–81, 84, 85, 89, 106, 113, 124, 127, 128, 129, 136, 137, 139, 143, 144, 145
Maryport (Cumbria), 74, 91
Meare Heath (Somerset), pollen count from, 105
Mellars, P. A., cited, 83
Meols (Lancs.): 93, 95–7; Iron Age and Roman finds from, 95
Mere Sands (Lancs.), occupation evidence at, 91
Mersey estuary (Lancs.), 74
*Midfendic*, the, 59, 61, 64, 65
Midley (Kent), 47
Midley Sand (Kent), 42
Milefortlets, 89, 91
Molenaarsgraaf (Netherlands), Neolithic settlements at, 118
Montrose (Tayside), 81
Morecambe Bay (Lancs.), 67, 74, 75, 81, 83, 91; *see also* Storrs Moss
Moricambe (Cumbria), 89, 91
Mörner, N. A., cited, 6, 8
Morrison, A., *see* Jardine, W. G.

Nancy's Bay (Lancs.), 81
Nash, S. G., cited, 26
Nennius, cited, 48–9
Netherlands coast: 106–31; dune-building in, 79, 112, 122, 123–4; geology, 107–110, 113, 119, 123, 130; palaeobotany, 106–7, 116–18, 123, 127; sea-level changes, 81, 107, 112, 113–15, 116, 122; Mesolithic period, 110, 115–16; Neolithic period, 110, 112, 115, 116–21, 123; Bronze Age period, 84, 110, 112–13, 123, 124–6; Iron Age period, 84, 110, 112, 123, 127–9; Roman period, 112, 122, 128, 129–30; Medieval period, 129, 131
Newchurch (Kent), 47, 48
Newlyn (Cornwall), 16, 17
New York harbour, 6
Nichols, H., cited, 87
North Downs, 12
North Sea: 12, 15, 17, 20; mesolithic finds from, 115
North Somerset Archaeological Research Group, 25, 26

Ordnance Survey: levelling networks of, 17, 21; tidal observatory of, 17
Oostwoud (Netherlands), beaker settlement at, 126

Pacific Ocean, 6, 8, 10
Palaeoenvironments: Lancashire, S.W., 76–80; Netherlands, 106–7, 116–18, 123, 127
Paterson, W. S. B., cited, 7
Peat, 1, 2, 15, 42, 45, 56, 69, 75, 78, 79, 87, 103, 104, 106, 107, 121, 122, 136, 137, 140
Peterston (Gwent), flood plaque in church at, 30
Pleistocene sea-levels, 1, 3, 7, 8, 10, 12, 16
Pentre (Clwyd), Roman lead industry at, 97
Pollen analysis, 76, 78, 80, 83, 84, 87, 91, 103, 104, 105, 106, 115, 119, 131, 137, 139, 140, 141, 142, 143, 144
Prestatyn (Clwyd): Roman mining centre at, 93, 97–101; medieval motte, 99
Ptolemy, cited, 69

Reade, T. M., cited, 75, 85
Radiocarbon dating, 42, 76, 79, 89, 105, 107, 115, 119, 123, 137, 138, 139, 142, 144
Rhee Wall, Romney Marsh (Kent), 37, 39, 49, 50
Rhine/Meuse delta (Netherlands), 110, 112, 118, 129
Ribble estuary (Lancs.), 74
Ribchester (Lancs.), 84, 102
Robert of Gloucester, Count, *see* Urban, Bishop of Llandaff
Robinson, A. H. W., cited, 15
Romney Marsh (Kent/E. Sussex): evolution of, 37–54; bibliography, 54–5; development processes, 39–40; Neolithic and Bronze Age afforestation, 42–3, 54; Roman and early Saxon, 43–7, 54; late Saxon, 47–9; medieval, 49–52
Romney, New, and Old (Kent), 46, 49, 50, 51
Rossall (Lancs.), 78
Rossiter, J. R., cited, 21
Rother, river, 39, 43, 45, 46, 49, 50, 51, 52
Rotterdam (Netherlands), Europoort, mesolithic finds froms, 115–16
Ruckinge (Kent), 47
Rufford (Lancs.), 91, 93
Rumney Great Warth (Gwent), erosion and redisposition of shards at, 25–6, 27

St. Mary-in-the Marsh (Kent), Roman occupation evidence at, 43–4
St. Woolo's Cathedral, *see* Gwynllyw, St.
Salinity changes, 10, 146
Salterns: Bronze Age, 69; Iron Age, 61; Roman, 44, 59, 61, 64, 65, 69
*Sandtun* (Kent), 49
Sea-level changes: 1–23, (Bibliography) 21–3; global influences, 6–10; instrumental influences, 2, 16, 17; local influences, 15; mean sea-level, or geoid, 3–6, 7, 10, 17, 20, 21; regional influences, 10–15; Lancashire, S.W., 75, 76, 78, 80–4; Netherlands, 107, 113–15; Somerset levels, 103–5
Sediments, 1, 2, 10, 15, 40, 46, 74, 76, 78, 79, 80, 84, 100, 105, 106, 107, 136–7
Severn Levels, 24–34; *see also* Gwent Levels
Sheerness (Kent), 17, 21, 136
Sibsey (Lincs.), 65
Simmons, Brian, 'Iron Age and Roman Coasts around the Wash', by, 56–73
Skegness (Lincs.), 67
Skippool (Lancs.), 84
Slea, river, 64, 71
Sleaford (Lincs.), 69
Smith, Charles Roach, cited, 37
Smith, D. E. *et al.*, cited, 81
Soil Survey of England and Wales, 53, 58, 69, 71
Solway Firth: coastal change in, 87, 89; Mesolithic period, 89; Neolithic and Early Bronze Age period, 89; Iron Age period, 89; Roman coastal defence, 89, 91
Somerset Levels: drainage of, 32; palaeobotany of, 103–5; sea-level change in, 103–5; Roman period in, 25
Speed, John, map of Lancashire by, cited, 91
Starr Hills (Lancs.), 79
Statorius Maximus, work at Caerleon by, 28
Sticklepath (Devon), earthquake at, 16

Stone Marsh (Kent), 142, 143
Storm pressure, 2, 12, 15, 17, 80, 134, 136, 145
Storrs Moss (Lancs.), 89
Streif, H., *see* Geyhl, M. A.
Sudbury (Suffolk), 17
Suthons, C. T., *see* Gordon, D. L.
Swifterbant (Netherlands): Mesolithic occupation at, 116; Neolithic settlements at, 112, 116–18
Swineshead (Lincs.), 61

Talbot, J. H., *see* Binney, E. W.
Tectonic movements, 2, 3, 8, 16, 136
Teichman Derville, M , cited, 38
Thames estuary: post-glacial environmental change and man in, 134–46, (bibliography 146–8); fauna, 144; human impact, 141–2, 143, 146; pollen analysis, 137, 139, 140, 141, 142, 143, 144; Mesolithic period, 141; Neolithic period, 141, 142; Bronze Age period, 144; Roman period, 145
'Theories of Coastal Change in North-West England', by Dr. M. J. Tooley, 74–86; bibliography, 85–6
Thomas, D. K., *see* George, K. J.
Thomas, G. S. P., cited, 83
Thurrock, West (Essex), 81
Tidal currents, 10, 15, 74, 80
Tide gauge records, 3, 7, 12, 17, 20, 21
*Terpen*, 113, 127–9
Tilbury (Essex): 81, 137, 138, 139, 140, 145; human skeleton from, 144
Tillingham, river, 39, 43, 45, 48
Tintern Abbey (Gwent), drainage of Moor Grange, Magor, by, 34
Tooley, Dr. M. J.: 'Theories of Coastal Change in North-West England', by, 74–86; cited, 40, 67, 91

Umbgrove, J. H. F., cited, 10
Unified European Levelling Network, 17
Urban, Bishop of Llandaff, and Count Robert of Gloucester, land use agreement between, 33

Van der Plassche, O., cited, 115
Van Regteren Altena, J. F., cited, 123
Velsen-Hoogovens (Netherlands), 79, 123
Vlaardingen culture, 110, 112, 119, 123

Walcott, R. I., cited, 8, 12
Walland Marsh (Kent/E. Sussex), 37, 40, 47, 51
Wallenberg, J. K., cited, 38
Walter, Earl of Pembroke and Lord of Striguil, 34
Ward, G., cited, 38, 39, 47
Wash: Iron Age and Roman coasts around, 56–73; for details *see* Lincolnshire Fens; Wash Barrage, 58
Waterbolk, H. T., cited, 127
Wemberham, Yatton (Avon), Roman villa at, 26, 34
Wemelsfelder, P. J., cited, 6
Westfrisia (Netherlands), Bronze Age settlements in, 110, 112, 113, 124–6
Whaplode (Lincs.), 61, 69
Williams, Michael, cited, 32
Winchelsea (E. Sussex), 37, 50
Wirral (Lancs.): 83, 95; finds from, Iron Age to Saxon, 95, 97
Witham, river, 65, 67, 71
Wood, J. G., cited, 34
Woodhall Spa (Lincs.), 71
Wrangle (Lincs.), 61
Wright, R. P., cited, 28